S0-BPC-173

Strategic Minerals

Strategic Minerals:

U.S. Alternatives

KENNETH A. KESSEL

NATIONAL DEFENSE UNIVERSITY PRESS
WASHINGTON, DC

National Defense University Press Publications. To increase general knowledge and inform discussion, NDU Press publishes books on subjects relating to US national security. Each year in this effort, The National Defense University, through the Institute for National Strategic Studies, hosts about two dozen Senior Fellows who engage in original research on national security issues. NDU Press publishes the best of this research. In addition, the Press publishes especially timely or distinguished writing on national security from authors outside the University, new editions of out-of-print defense classics, and books based on conferences concerning national security affairs.

NDU Press publications are sold by the US Government Printing Office. For ordering information, call (202) 783-3238, or write to: Superintendent of Documents, US Government Printing Office, Washington, DC 20402.

William A Palmer, Jr., Cheltenham, MD, proofread this book under contract DAHC32-87-A-0015.

Library of Congress Cataloging-in-Publication Data

Kessel, Kenneth A.
Strategic minerals : US alternatives / Kenneth A. Kessel.
p. cm.
Includes bibliographical references.
$10.00 (est.)
1. Mineral industries—Government policy—United States. 2. Strategic minerals—Government policy—United States. 3. United States—National security. I. Title.
HD9506.U62K47 1990
333.8'5'0973—dc20 89-23071
CIP

First printing, February 1990

To my parents, to my brother and sister,
and especially to my son, Ryan.

Contents

Illustrations

FIGURES

Foreword

ENSURING ADEQUATE SUPPLIES OF STRATEGIC MINERALS has long been a national concern. Congress held hearings on this problem as far back as 1880. In 1939, the Government established a critical minerals stockpile. Yet, as author Kenneth Kessel points out, the country is still in search of a comprehensive policy. Too often, this search has been buffeted—sometimes blown off course—by the unpredictable winds of politics, economics, international crises—and flawed perceptions.

Many studies of the issue conclude that because of a mineral dependency—often caused by ill-advised manipulation of the national stockpile—the United States suffers strategic vulnerability. But dependency, Kessel contends, does not necessarily mean vulnerability. He argues that such pessimistic perceptions are based on unrealistic assumptions and faulty premises—notably, the idea that certain minerals could not be obtained at all. *Legitimate* worst case scenarios, Kessel says, do not place the US and its allies in such dire straits. Although peacetime economics makes cheaper foreign sources more attractive than domestic ones, other approaches—synthetics, recycling, substitution—and other sources can be tapped when necessary. New technology, furthermore, continues to offer alternatives to strategic minerals.

Author Kessel thus advises us to put to rest unwarranted concerns about the minerals vulnerability issue. As reassurance, he proposes better ways to identify and obtain materials for the national stockpile. Supported by a wealth of data and thought-provoking analysis, this study challenges conventional approaches and offers reasonable alternatives for US critical minerals policy.

J. A. BALDWIN
VICE ADMIRAL, US NAVY
PRESIDENT, NATIONAL DEFENSE UNIVERSITY

Strategic and Critical Materials Chronology

Major Direct Legislative Acts

1922	Army and Navy Munitions Board Created
1939	Strategic and Critical Stock Piling Act
1946	Strategic and Critical Materials Stock Piling Act
1950	Defense Production Act
1970	Resource Recovery Act
	National Materials Policy Act
	Mining and Minerals Policy Act
1974	National Commission on Supplies and Shortages Act
1979	Strategic and Critical Materials Stock Piling Revision Act
	Deep Seabed Hard Minerals Resources Act
1980	National Materials and Minerals Policy, Research and Development Act
1984	National Critical Materials Act

Related Legislative Acts

1934	US Export-Import Bank
1949	Commodity Credit Corporation Charter Act
1954	Agricultural Trade Development and Assistance Act
1961	Agency for International Development

Note: This list is not exhaustive. The Departments of State and Commerce, DOD, DIA, and CIA have also completed additional studies. This list is intended to capture the history and extensiveness of the work that has gone into the study of strategic and critical materials.

1969 Overseas Private Investment Corporation
1980 IDCA Trade and Development Program

Presidential Initiatives

1949 Hoover Commission created the Department of National Resources
1958 Eisenhower Strategic and Critical Materials Study reduced goals from 5 years to 3 years
Kennedy National Materials Management Review
1973 Nixon Stockpile Study reduced goals from 3 years to 1 year
1977 Carter Nonfuel Minerals Policy Review
1982 Reagan National Materials and Minerals Program Plan and Report to Congress
1983 Reagan creation of Office of Strategic Resources within the Department of Commerce
1984 Reagan creation of National Strategic Materials and Minerals Program Advisory Committee with the Department of Interior
Reagan creation of the Cabinet-level National Critical Materials Council
1985 Reagan Revised Stockpile Recommendation
1988 Reagan Executive Order names DOD as National Defense Stockpile Manager

Reports, Studies, Conferences and Commissions

1951 Truman Materials Policy Commission (Paley Commission)
1970 Congressional National Commission on Materials Policy
1974 National Commission on Supplies and Shortages
1976 National Security Council Stockpile Review Studies
1984 National Security Council Stockpile Review Studies

1986 National Strategic Materials and Minerals Program Advisory Committee (Mott Report)

Office of Technology Assessment

1976 Assessment of Information Systems Capabilities Required to Support US Materials Policies Decisions

Assessment of Alternative Economic Stockpiling Policies

1979 Management of Fuel and Nonfuel Minerals on Federal Lands

Technical Options for Conservation of Metals

1985 Strategic Materials: Technologies to Reduce US Import Vulnerability

US General Accounting Office

1960 Review of Procedures and Practices of General Services Administration (GSA) Relating to the Storage and Physical Inventory of Strategic and Critical Materials at GSA Storage Depots.

1964 Unnecessary Cost Incurred by the Government by Not Using Surplus Stockpiled Materials to Satisfy Defense Contract Needs—Department of the Air Force

1971 Management of Selected Aspects of the Strategic and Critical Stockpile

1974 US Actions Needed to Cope with Commodity Shortages

1976 US Dependence on Five Critical Minerals

1978 Deep Ocean Mining

The Department of Interior's Minerals Availability System

Interior Programs for Assessing Mineral Resources on Federal Lands Need Improvements and Acceleration

1979 Mining Law Reform and Balanced Resource Management

Learning to Look Ahead: The Need for a National Materials Policy and Planning Process

Phosphates: A Case Study of a Valuable, Depleting Mineral in America

1980 Industrial Wastes: An Unexplored Source of Valuable Minerals

1981 Minerals Management at the Department of Interior Needs Coordination and Organization

1982 Actions Needed to Promote a Stable Supply of Strategic and Critical Minerals and Materials

1986 National Defense Stockpile: Adequacy of National Security Council Study for Setting Stockpile Goals

1987 National Defense Stockpile: National Security Council Study Inadequate to set Stockpile Goals

Congressional Research Service (US Library of Congress)

1942 Raw Materials (April 1941-March 1942) Selected and Annotated Bibliography on Raw Materials in a Wartime Economy

1972 Resolving Some Selected Issues of a National Materials Policy

1976 Materials Availability Bibliography

Ocean Manganese Nodules

1981 A Congressional Handbook on US Materials Import Dependency/Vulnerability

1986 The Reagan Administration Proposes Dramatic Changes to National Defense Stockpile Goals

National Defense Stockpile Policy—The Congressional Debate

US Congressional Committees

1969 Senate Committee on Public Works Towards a National Materials Policy

1976 Joint Committee on Defense Production: Federal Materials Policy Recommendations for Action

1977 House Committee on Science and Technology: Materials Policy Handbook: Legislative Issues of Materials Research and Technology

1978 House Committee on Science and Technology: Building a Consensus on Legislation for a National Material Policy

1980 Senate Committee on Foreign Relations: Imports of Minerals from South Africa by the United States and the OECD Countries

1981 House Committee on Science and Technology: Innovation in the Basic Materials Industries

National Academy of Sciences

1975 National Materials Policy

1978 National Materials Advisory Board (NMAB)—Contingency Planning For Chromium Utilization

1985 NMAB—Basic and Strategic Metal Industries: Threats and Opportunities

Bicentennial National Materials Policy Conferences

1970-86 Sponsored by the Engineering Foundation and the Federation of Materials Societies

Department of Interior
Office of Minerals Policy Analysis

1981 Cobalt: Effectiveness of US Policies to Reduce the Costs of a Supply Disruption

Chromium: Effectiveness of US Policies to Reduce the Costs of a Supply Disruption

1982 Manganese: Effectiveness of US Policies to Reduce the Costs of a Supply Disruption

Platinum and Palladium: Effectiveness of US Policies to Reduce the Costs of a Supply Disruption

1985 Bureau of Mines (BOM) Research

1986 South Africa and Critical Materials
1987 BOM Mineral Commodity Summaries

Bureau of Mines Research

1985 Inventory of Land Use Restraints Program
Minerals Availability Appraisals for Asbestos, Cobalt, Fluorspar, Mercury, Molybdenum, Lead, Zinc, Titanium, and Tungsten
Mineral Industry Surveys of Supply—Demand for Ferrous Metals, Non-ferrous Metals, and Industrial Minerals
Chromium Industry of the USSR
Recovery of Chromium, Cobalt, and Nickel from Mixed and Contaminated Superalloy Scrap
Consolidation and Joining of Rapidly Solidified Alloy
Gallium and Germanium from Domestic Resources
Characterization of Ocean Floor Minerals
Improved Precious Metal Recovery
Chromium—Free Heat Treatable Alloy Steel
Hardfacing Alloys Containing Minimal Critical Metals
Probabilistic Assessment of Undiscovered Mineral Resources in Alaska
Recovery of Critical Metals from Waste Catalysts

National Defense University

1978 Managing Critical and Strategic Non-Fuel Minerals since World War II
1984 Industrial College of the Armed Forces (ICAF): The South African Ferrochromium Industry and Its Implications for the US
1985 Industrial Preparedness: Breaking with an Erratic Past
Mobilization and the National Defense

Federal Emergency Management Agency

Annual Stockpile Reports to Congress
Annual Materials Plans

Other Agencies and Departments

1973 NATO Classified Study on Strategic Mineral Dependence of the NATO Countries

1983 US Strategic Minerals Policy: The Significance of Southern African Sources (Prepared by Dr. Stephen Wergert while on sabbatical from the Defense Intelligence Agency)

Review of 21 Superalloys and Advanced Materials (US Air Force ongoing study)

1985 Department of Commerce: US Dependence on South Africa for Strategic Materials (Classified)

South Africa—A United States Policy Dilemma (Air War College)

1987 Department of State: Report on South African Imports, transmitted to Congress on 30 January 1987 (in response to Section 504(a) of the Comprchcnsive Anti-Apartheid Act of 1986—Public Law 99-4440)

Strategic Minerals

1

Mineral Dependency: The Vulnerability Issue

THE STRATEGIC MINERAL DEPENDENCY OF THE UNITED STATES is not a recent development. In fact, US dependency has a long history. But no country in the world is totally self-sufficient in minerals, not even the USSR, despite its long-standing policy of economic self-sufficiency. The *use* of the word "dependent" does imply a vulnerability, however, and that is the crux of some concern. A perhaps more accurate description of the status of US strategic minerals is "domestic insufficiency" at current market prices or "foreign trade deficit." Nonetheless, the terms dependency and vulnerability are in common use and form the basis of the first two chapters of this book. The first chapter surveys the pertinent data on global mineral resources, production, and trade among the major players—comparing, in particular, the degree of import dependence among them and defining "strategic minerals."

Resources in the United States

The United States is heavily dependent on imports of key minerals used in the production of strategic items such as military jet engines, avionics, ships and tanks, artillery, and space vehicles. Although this is a source of current concern for policy-makers, it is not a recent development. Historically, the United States has never been self-sufficient in either strategic or nonstrategic minerals. Until the Great Depression, however, a balance between mineral imports and exports was maintained. During the period 1900-1929 the United States produced nearly 90 percent of all the minerals it consumed. Since about World War II, however, the mineral position of the United States has deteriorated. Rising nationalism in the Third World, accompanied by nationalization, expropriation, increased taxation, and constraints on the degree of foreign ownership have since limited control of foreign minerals by US companies. Moreover, sharply higher energy costs since 1973 and a trend in the less developed countries toward processing their own ores have led to a decline not only in mining but also in the US mineral processing industry. Restrictive environmental regulations have made domestic mineral exploitation and processing more difficult and costly than in the past and have greatly increased lead times for new mineral ventures in the United States. For example, according to a major copper company, it is subject to rules and regulations issued by 54 different federal departments and agencies, 42 state boards and commissions, and 39 local government units, raising cumulative production costs of a pound of copper by 10 to 15.1 cents.[1]

Despite such developments, imported mineral raw materials of some $4 billion constitute less than 15 percent of total US mineral raw material requirements. In fact, the value of such imports equals only the value of recycled and reclaimed mineral materials.[2] However, when processed materials such as steel, aluminum, and ferroalloys are included, the US mineral trade balance shows a $13 billion deficit, based on imports of $37 billion and exports of only $24 billion. Putting these numbers in perspective, the total value of processed materials of mineral origin is about $240 billion or slightly less than 6 percent of the US Gross National Product. This low percentage reflects the maturity of the US economy from one based on "smokestack" industries to one dominated by services and technology. In order to reach this point in economic development, however, minerals have been crucial, as seen in historical consumption data. Estimates show that in the last 35 years alone, the United States has consumed more minerals than did all mankind from the beginning of time until about 1940.[3]

With just over 5 percent of the world's population, the United States still consumes about 23 percent of the world's use of nonfuel mineral resources for a per capita average of about 22,000 pounds per year. No other country comes close to this per capita rate of use.

The United States, moreover, did not become one of the world's leading industrial powers by being a mineral-short nation. It has economic reserves in all but 12 of the 71 minerals in world demand and in many of those 12—e.g., chromium, cobalt, and manganese—it has resources that, while uneconomic at today's low prices, could be produced at higher prices (appendix A, table A-1). Indeed, the United States mines and processes more than 90 metals and nonmetals. Much of this production, however, is concentrated among the 20 highest valued minerals, which account for over 90 percent of the value of production.[4] The United States is especially rich in the nonmetallic minerals, producing in value terms nearly three times the amount of its metallic mineral output. The visibility of its strategic mineral imports also overshadows the fact that the United States exports more than 60 different metallic and nonmetallic minerals. In sum, the United States ranks first or second in the world in the production of nonfuel minerals, the rank depending on the level of overall economic activity.

Concentration of Global Supplies

Much of the world's mineral wealth is located outside the United States. This generalization holds especially true for the strategic minerals. In many cases, foreign ore deposits are richer than those found in the United States, are located close to cheap energy sources, and are mined by low-cost labor. Moreover, industrial demand in these mineral-rich countries is considerably lower than the available supplies, making them ideal export commodities. In effect, economic factors abroad have played an important role in contributing to the current state of US foreign mineral dependence.

After the United States, the other major producers—ranked in order of value of mineral production—are the USSR, Canada, Australia, and South Africa. Of these four, the *USSR* ranks first in the world in the production of iron ore, manganese, steel, and platinum-group metals and second in aluminum, lead, nickel, and gold. Once a major force in world metals markets as an exporter, the importance of the USSR—especially its strategic minerals—has declined as its own internal demands have caught up to its domestic mining capacity. Although there is no shortage of domestic mineral resources within the Soviet Union, exploration and development costs have risen sharply since the 1950s, with the richest and easiest-to-mine ores exploited first. Because of the sparse data released to the West, no one knows whether Soviet mine capacity and production can be maintained in the face of rising industrial demand. It is clear, however, that the production base is moving to more inhospitable regions of the USSR, where ores are less accessible. This geographic trend in Soviet mining capacity forms the basis of one hypothesis of a US security concern—the "Resource War," which I expand on in chapter 2. The gist and premise of this particular concern suggests that it may be cheaper for the USSR to compete with the West in world metals markets than to further develop its own resources.

Among the most critical strategic minerals, the USSR currently is a major exporter of only one—the platinum group—despite being at or near the top in world resources and production of all of the others (appendix A, tables A-2 through A-4). Among the platinum-group metals (PGMs), the USSR is the world's largest palladium producer at 1.8 million troy ounces. Its platinum production of 1.5 million troy ounces ranks a distant second behind South Africa. Palladium exports to the West of 1.2 million troy ounces account for more than 50 percent of total consumption. Platinum exports are much lower—less than 400,000 ounces in 1987 or only about 10 percent of Western consumption. The PGM export mix reflects a relative shortage of platinum in Soviet ores, which are found in a palladium-platinum ratio of about 5:1. The USSR is also a significant exporter of rhodium, required in today's three-way automobile catalytic converters.

Although the USSR is the world's largest manganese producer, it has not exported any manganese to the West since 1978. Small quantities of manganese concentrate and ferromanganese are exported to Eastern Europe. The USSR has itself become increasingly dependent on the West for high-quality manganese ore, importing several hundred thousand tons from Gabon, Australia, and Brazil. A Soviet foreign trade official has indicated that imports are being driven by a depletion in traditional ores, higher demand, and increased requirements for higher quality materials.[5] It seems unlikely that the USSR will become a major supplier of manganese to the West until its steel industry is modernized and it becomes more efficient in manganese use per ton of steel produced.

The USSR also was once a major source of chromium for the United States—supplying in 1970 nearly 60 percent of US imports of metallurgical grade chromium at the height of the 1966-72 United Nations' embargo on Rhodesian chromium exports. Subsequently, the Soviet share of all types of US chromium imports fell to only 8 percent by 1981 as US ability to process lower grade South African ore increased and as demand shifted from basic ore to ferrochromium. By 1983 all chromium exports to the United States had ceased. USSR chrome exports resumed in small quantities in 1986—apparently because of US concern over supplies from South Africa. Japan was once a major purchaser of Soviet chromium ore, but purchases have fallen to about 50,000 tons per year. At their peak in 1970, Soviet chromium exports to the West totaled one million tons. The USSR still exports very small quantities of ferrochromium—about 10 to 15 thousand tons per year, mostly to Western Europe.[6]

For cobalt, the USSR is a net importer, depending on Cuba for 45 percent of its consumption needs and importing a small amount from Zaire. A major expansion in cobalt production from a rich ore deposit at the Noril'sk mining complex in eastern Siberia will reportedly increase output there by one-third, but this will not be enough to allow the USSR to export cobalt. It could, however, free up some supplies from Cuba for export to the West for hard currency. Little is known about Soviet vanadium production. The Bureau of Mines estimated that output in 1984 was about 9,500 metric tons or 25 percent of total world output. Small quantities of vanadium pentoxide (used to produce ferrovanadium) are imported from Finland and some vanadium slag and ferrovanadium apparently are sent to Eastern Europe.

Canada. Canada produces more than 60 mineral commodities, ranking first in the world in the production of nickel, zinc, potash, and asbestos and second in the production of molybdenum and uranium. More important, perhaps, is that Canada exports about two-thirds of its nonfuel mineral output, with some 70 percent of such exports going to the United States in the form of asbestos, potash, gypsum, iron ore, nickel, silver, zinc, copper, and lead. In strategic minerals, Canada is a small but nonetheless important producer of several—accounting for 15 percent of the world's production of columbium, 9 percent of all cobalt, 4 percent of all platinum-group metals, and 2.5 percent of all tungsten. In 1985 Canada accounted for 7 percent of US platinum imports. Over the longer term, these supplies could increase as a result of promising exploration projects now underway.

Australia. Australia produces more than 70 minerals, leading the world in the output of bauxite, alumina, rutile, and ilmenite concentrates—ores and intermediate products necessary for the production of aluminum and titanium. In trade Australia's mineral exports account for over one-third of its total exports. Australia ranks first, second, or third in the export of alumina, iron ore, lead ilmenite, rutile, zinc, bauxite, nickel, and tungsten. Other important exports include manganese, copper, tin, and silver. Moreover, Australian mineral exploration and development are far outpacing its industrial economic growth so that this country should become an ever more important source of minerals to the West in coming years.

South Africa. The Republic of South Africa deserves special attention as the richest source of strategic minerals in the world. Geologic fate was extremely kind to the country in two ways. As is the case with Canada, Brazil, Australia, and the USSR, South Africa possesses the world's oldest (Archaean) rock—rich in gold, iron and manganese.[7] Moreover South Africa's rock is exposed and easily accessible to mineral development, where Brazil's formations are covered by rain forests, Canada's by glacial debris, Australia's with the thick soils of the outback, and the USSR's with steppes. And, about 1.8 billion years ago the world's largest single igneous rock body was formed—about the size of Ireland—

as a result of continental drifting, leaving behind huge deposits of platinum, chromium, and diamonds.

The statistics for South Africa confirm its massive mineral wealth. In chromium, South Africa contains 85 percent of the world's proven reserves and over 70 percent of its chromium resources; in manganese, the equivalent percentages are 70 and 75; in the platinum-group metals, 90 percent of the world's reserves; for vanadium, about 50 percent of reserves.[8]

As one might expect, the mineral production picture in South Africa parallels the richness of its mineral endowment. It is the world's largest producer of gold and vanadium; second largest producer of chromium, manganese, and platinum-group metals; and third largest producer of industrial diamonds. With the exception of industrial diamonds and palladium, South Africa is the largest exporter of all of these minerals; based on historical averages, it has provided 5 percent or more of US consumption of nine major mineral commodities, seven of which appear on the national stockpile list of strategic and critical materials.[9] In several instances, these shares are many times higher than 5 percent. South Africa is the only source of large-particle, coarse-grade andalusite—critical for making the refractories that are used to line blast furnaces used for iron and steelmaking. South Africa also supplies about 85 percent of US consumption of platinum-group metals; 60 percent of US consumption of chrome ore; and 47 percent of US consumption of ferrochromium. It also provides 35 percent of US imports of manganese ferroalloys. Until about 1986-87, Pretoria was considered to be an extremely reliable minerals supplier, but the worsening of political conditions inside the country has forced political-risk analysts to reassess the situation.

A 1985 US commodity study brought home the across-the-board dominance of the world's top five mineral producers (appendix A, Table A-5). More than 50 percent of the world's production of 10 of 25 important minerals originates with the United States, Canada, Australia, South Africa, and the Soviet Union.[10] The USSR and South Africa alone produce more than 50 percent of world production for five of them. Of these five, it is nearly universally agreed that three—platinum, manganese, and chromium—are of strategic concern to the West because of their important military and essential civilian uses.

Zaire and Zambia. Among the Less Developed Countries (LDCs), Zaire and Zambia merit special mention because of their importance as world cobalt suppliers. Both produce cobalt as a by-product of copper production from a rich copper-belt extending roughly along their mutual border into the southern region of Zaire known as Shaba. Zaire's ores have one of the world's highest cobalt concentrations at .3 percent, while cobalt concentrations in Zambia are about half this percentage.[11] In terms of reserves, Zaire ranks first in the world with 25 percent of the world's total and Zambia sixth. In terms of production, Zaire accounts for nearly 60 percent of the world total. Zambia is second in world output, producing 11 percent. Moreover, these percentages are understated because about 13 percent of global output is produced by Cuba and the USSR and consumed within the Soviet Bloc. From an export standpoint, Zaire and Zambia together supply 53 percent of US cobalt imports and nearly 50 percent of annual consumption. Many experts do not consider these supplies to be reliable because at least 50 percent of these exports must pass through South Africa by truck or rail to the ports of Port Elizabeth or East London. Some alternate supply possibilities exist, however, and these are discussed chapter 2.

The rest of the world. Surprisingly, the high degree of concentration of mineral reserves that exists for the world's top five producers extends to the Less Developed Countries as well. (See figure 1.) Fewer than 30 LDCs have reserves as high as 5 percent of the world's total for 25 different minerals. The LDC reserve pattern differs most in that only one major mineral commodity is usually found in a given LDC in significant quantity.[12] Among the major exceptions—meaning countries with more than one major mineral resource—are Brazil (columbium, manganese, and tantalum), Peru and Mexico (silver, zinc, and copper), Thailand (columbium, tin, tantalum), Zaire and Zambia (copper and cobalt), and the PRC (multiple minerals). In some cases, major new mineral discoveries could well be added to the list in large countries such as Brazil and China, countries where mineral exploration is in its relative infancy and where geologic formations are prone to mineral formation. In other cases such as Africa, deposits of new minerals could well be found in some countries where exploration has not been systematic. However, outside of southern Africa, new discoveries are likely to be small in world terms, although important to the individual country as an additional source of foreign exchange.

Abundant World Mineral Reserves

Some alarmistic groups have postulated that the world will soon run out of basic necessities, especially food and minerals, but the evidence does not support their case.[13] This misplaced alarm was especially prevalent during the first half of the 1970s when (a) several consecutive poor world wheat crops did indeed create a temporary shortage situation, (b) a series of strikes and natural disasters led to concern about the inadequacy of the world's mineral supplies, and (c) the 1973 Arab oil embargo created a near-panic situation. The dire calculations were erroneous because they were based on extrapolations of past growth trends in material supply and demand. Driving the calculations were linear projections of population growth, which created pressure from the demand side, and a failure to allow for technological change, creating pressure on the supply side. In the area of agriculture, for example, improved yields have since led to record global food surpluses, while near-zero population growth in several countries has pulled the rug from under population projections.

In the minerals area, an examination of Bureau of Mines estimates of reserves and cumulative demand to the year 2000 (table A-1) shows little cause for concern. The tightest situation is projected for asbestos, industrial diamonds, indium, sulfur, tin, and zinc. For the other 63 minerals listed, reserve estimates are several multiples of demand. Moreover, these reserve estimates are likely to be conservative.

Three factors could well lead to a continuation of surplus mineral availability well into the next century. First, the post-industrial age in the developed countries has meant that fewer raw materials are being used to generate a dollar of GNP. The ratio

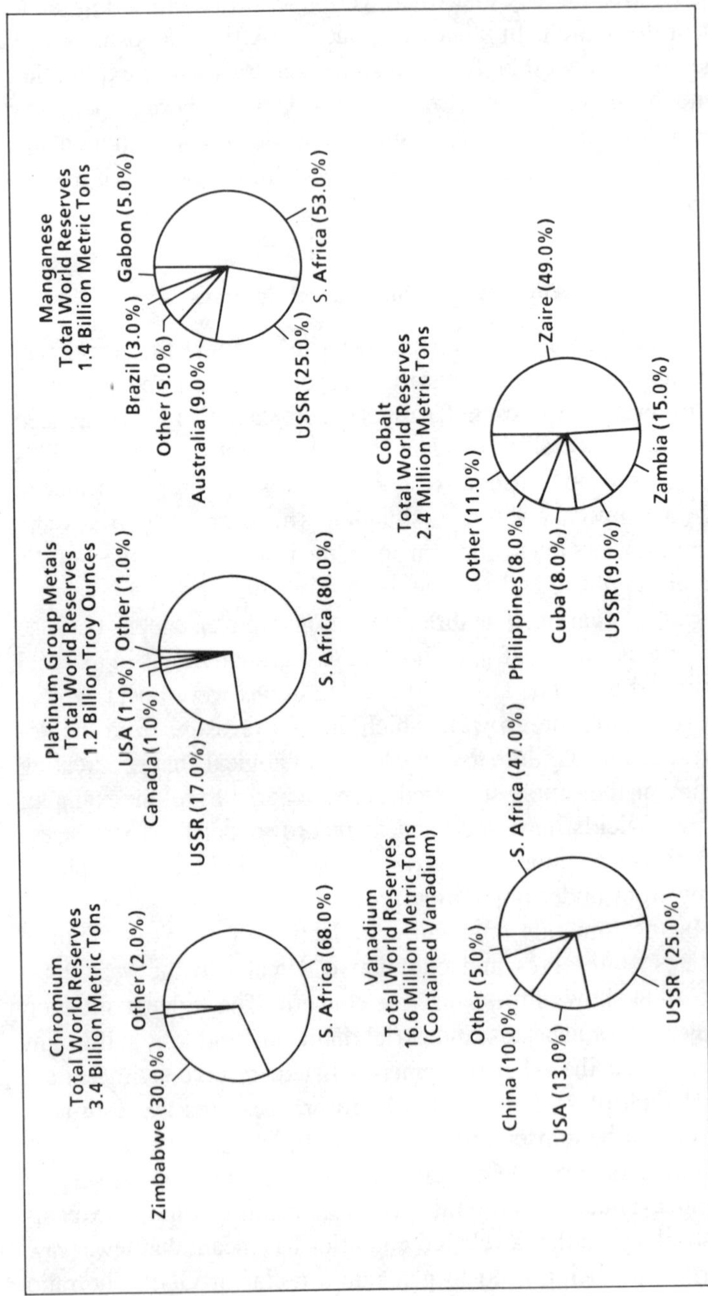

Figure 1. Estimated Percentage of Total World Reserves.

of minerals demand to GNP—called the intensity of use factor—is on the decline for most of the bulk metals consumed by the countries of the Organization for Economic Cooperation and Development (OECD). In the United States in 1972, for example, to generate one billion dollars of GNP required 1.6 tons of copper input. By 1984, the ratio fell to only .6 tons.[14] Second, the LDCs are entering a stage of slower overall economic growth, greatly offsetting a rise in their own intensity of use. Economic forecasters such as Wharton and others estimate that the 1980s will show LDC growth on the order of only 4 percent, compared with nearly 6 percent in the 1970s. Third, the LDCs are unlikely to mirror the industrial development stages experienced by most of the current OECD countries. By by-passing extensive growth in heavy industry for technology-dominated industries, many LDCs will never reach the high intensity-of-use levels experienced by the United States and Western Europe. Thus, it would seem that whatever US problems arise from a high degree of reliance on imports, they will not likely be compounded by any absolute shortages in world mineral supplies.

Defining Strategic Minerals

The focus of the remainder of this book will be the strategic minerals and their implications for US vulnerability, related policies, and the public good. Perhaps the most difficult single exercise is in simply identifying the "strategic" minerals. Lists abound and no two lists are alike. The first attempt to define strategic minerals was made by the Army and Navy Munitions Board following World War I.[15] Two classifications were identified—strategic materials and critical materials. Strategic materials were distinguished by their essentiality to the national defense, their high degree of import-reliance in wartime, and the need for strict conservation and distribution control. Critical materials were considered less essential and more available domestically, requiring some degree of conservation. Note the subjectivity of the definitions, especially for the critical materials. In 1944 the distinction between strategic and critical was abandoned and the definition simplified to "being essential in war" and requiring "prior provisioning" or stockpiling.

The current US definition, according to the Strategic and Critical Materials Stock-Piling Act of 1979, as amended, defines strategic and critical materials as those that (a) would be needed to supply the military, industrial, and essential civilian needs of the United States during a national emergency, and (b) are not found or produced in the United States in sufficient quantities to meet such need. The current definition differs little from the one established some 50 years prior. The first part of the definition implies essentiality. Figure 2 graphically illustrates US industrial uses for five key strategic minerals. The second part of the current definition relates to import reliance.

For operational purposes, in recommending and carrying out national stockpile policy and management, the Federal Emergency Management Agency (FEMA) defines "strategic" as the "relative availability" of a material and "critical" as its "essentiality." Whether the current National Defense Stockpile is consistent with the above definitions is open to question. Nonetheless, it currently consists of inventories of 74 distinguishable materials (appendix A, table A-6). Many of these 74 materials are various forms or grades of the same mineral. Depending upon how one groups these materials, the stockpile contains roughly 35 basic minerals, of which some 25 are metallic and the remainder nonmetallic.

A High Degree of Import Dependence

The word *strategic* has come to mean, more than anything else, "import dependent"—defined as the ratio of imports to consumption. Subsumed in this usage is a common understanding that the exporting countries are somewhat unreliable as a source of supplies. An examination of US Bureau of Mines dependence figures for 1985 for selected minerals (figure 3) indicates that imports constitute more than 50 percent of US consumption for 20 nonfuel minerals. In the 90 to 100 percent range are platinum, cobalt, and manganese—which are found primarily in southern Africa—as well as bauxite, columbium, mica, strontium, and

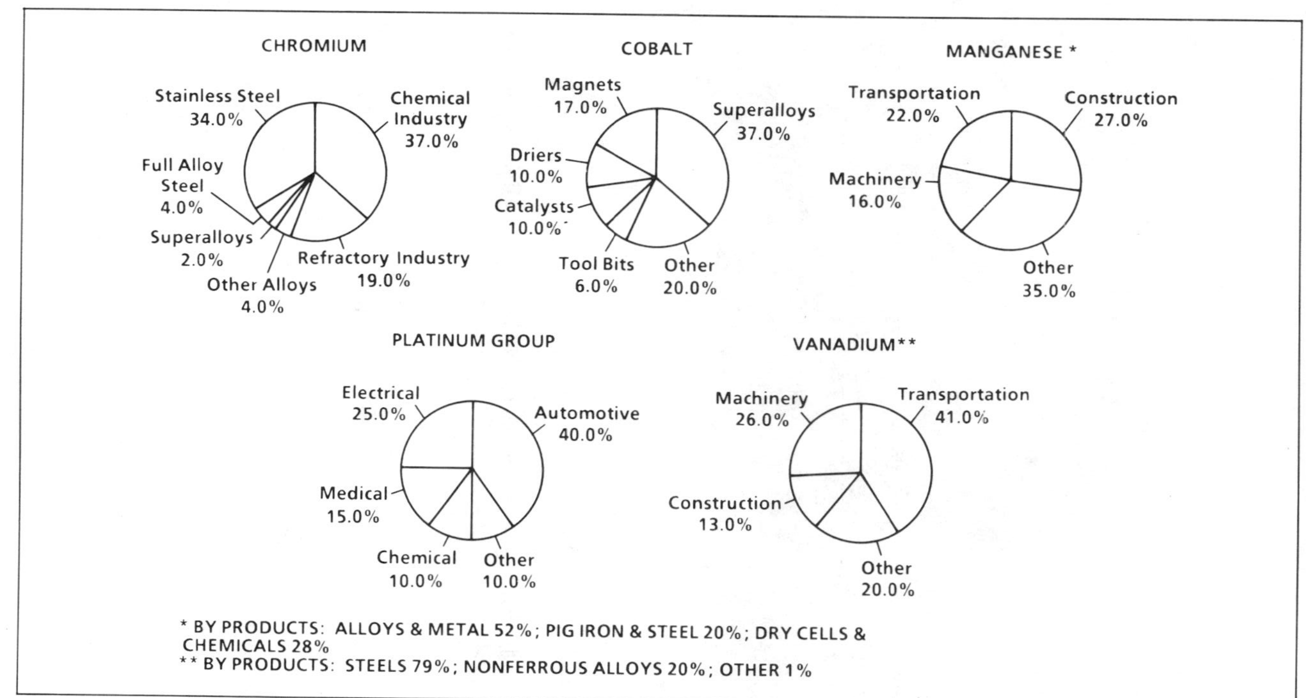

Figure 2. Key Minerals: US Consumption By End Use.

Figure 3. 1985 Net Import Reliance Selected Nonfuel Mineral Materials.

tantalum. In the case of chromium, the import dependence percentage has generally been lower—in the low- to mid-80s range—because of the high degree of recycling of stainless steel scrap.

Western Europe and particularly Japan are even less self-sufficient than the United States (figure 4). The European Community shows some 22 minerals in the 50 to 100 percent range and Japan has 24. Japan is more than 90 percent dependent for 18 minerals and 100 percent dependent for 10. A look at the lower half of figures 3 and 4 indicates that, for a wide variety of minerals, the United States is relatively well-off, compared to Europe and Japan.

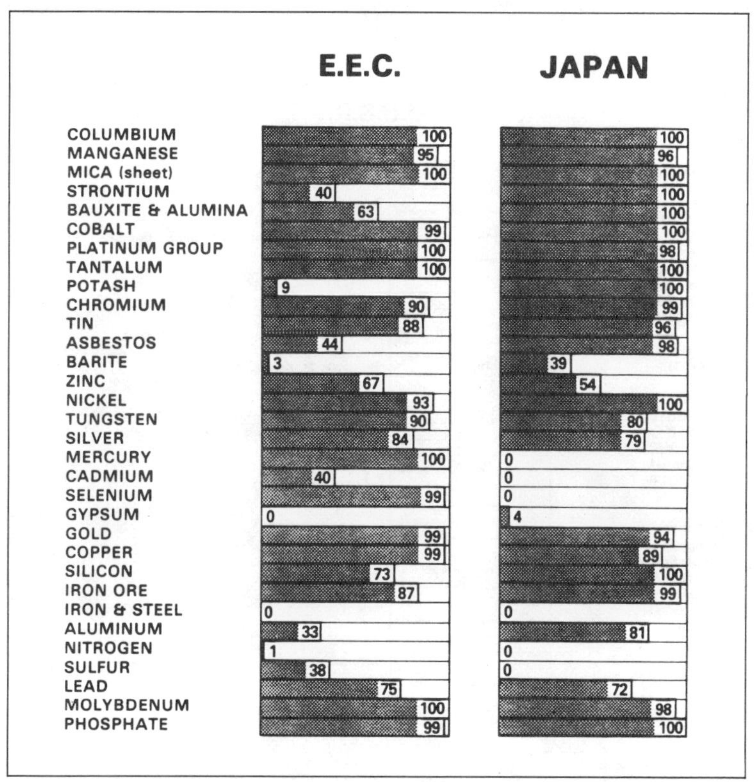

Figure 4. 1984 Net Import Reliance Selected Nonfuel Mineral Materials.

The USSR (figure 5) is in the best shape of all countries. According to the Bureau of Mines, the Soviets enjoy a situation of near-total self-sufficiency. In no case is the USSR as much as 50 percent dependent on imports and in only four cases are the dependence figures in the 40 percent range. Moreover, in three of the four cases—cobalt, bauxite, and tungsten—supplies come primarily from Cuba, Eastern Europe, China, and Mongolia. Only for bauxite and tin is the USSR relatively highly dependent on Western supplies. The USSR's high degree of self-sufficiency in nonfuel minerals is explained primarily by two factors: (a) with one-sixth of the world's land mass, it is richly endowed with

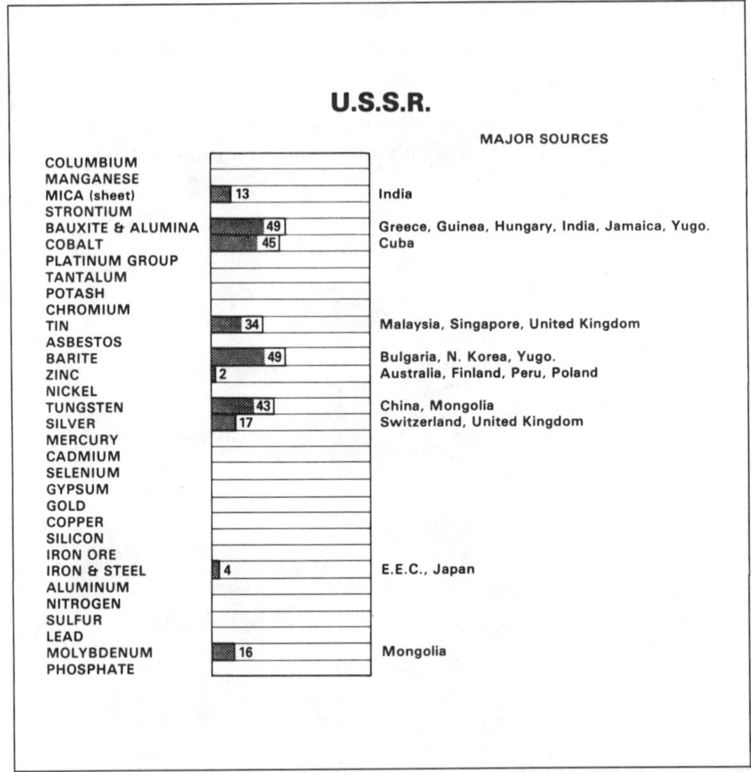

Figure 5. 1985 Net Import Reliance Selected Nonfuel Mineral Materials.

minerals; and (b) it has pursued a policy of autarky, i.e., the establishment of a self-sufficient and independent national economy. Moscow has not always been successful in this approach—insufficient grain production being one of its glaring failures—but for minerals, the policy has worked and worked well, albeit probably at great cost.

Mineral import dependence of the United States—while a security concern—has provided the greatest net benefit to the largest number of people. The costs of self-sufficiency are enormous and the benefits of free trade well established. Elementary trade theory is based on the law of comparative advantage, which in simplified terms says that if two countries each specialize only in goods which they can produce most efficiently, both sides will benefit by trading the goods that they produce for the goods produced by the other country, i.e., an economic "win-win" situation is created. This has been especially true in minerals and metals such as copper, aluminum, and steel—in which the United States has experienced great declines in production, with foreign suppliers sending cheaper products to the United States.

Benefits of Import Dependence

The benefits of interdependence in trade were supported by the National Commission on Supplies and Shortages in a 1974 report, which stated that the United States imports nonfuel minerals because the nation and its citizens gain economic advantages—lower costs and higher real incomes, i.e., the economic public good is well-served. The report also made an important distinction, noting that import dependence was more properly a minerals trade deficit and rejected the idea that this somehow represented a national loss. Indeed, the word *dependence* is an emotion-packed word and implies a certain degree of vulnerability; i.e., as a child is dependent on its parents, it is vulnerable in the absence of its parents' care.

Both the 1974 commission and a commission created by the National Materials Policy Act of 1970 noted, however, that there could be compelling national security reasons for forgoing the advantages of a mineral's trade-deficit situation. In its 1973 final report to the President and to the Congress, the National Materials commission concluded that

> in the interest of national security, it is unwise to become dependent upon specific strategic commodities for which the United States lacks a resource base and which are obtained mainly from a small number of countries which may choose to restrict or cut off the flow of supply . . . . The interest of national security will be served by maintaining access to a reasonable number of diverse suppliers for as many materials as possible.[16]

To reduce this import dependence, the committee recommended that the United States should attempt to foster the expansion of domestic production, diversify sources of supply, find substitutes or develop synthetics, increase the dependence of supplying countries upon continuing US goodwill, and allocate existing supplies through a priority use system. While these are all good and great goals, they do not come without a price. In some instances, such as expanding domestic production or developing substitutes, that price may be very high.

The "List of Ten." The Comprehensive Anti-Apartheid Act of 1986 required the administration to prepare three studies on strategic minerals. Responsibility for these studies was delegated to the Secretary of State, in consultation with the Departments of Commerce and Defense. Section 303 of the act prohibits the importation of most goods from South African parastatals. Under Section 303 (a)(2), exemptions are permitted for those strategic minerals that are essential to the economy or defense of the United States that are unavailable in sufficient quantities from other reliable and secure suppliers. Thus, the task of the first study was to certify a list of exempt minerals with justifications for their exemption. The creation of a working group to study the problem allowed for input by several agencies within the executive branch. In attempting to define a list of exempt items some participants believed the list should contain every item in the current National Defense Stockpile inventory. At the other extreme, some felt that nothing should be on the list because of the worldwide glut of almost every mineral. A modest list was eventually pared down to 10 items and a consensus achieved. The "List of Ten" is andalusite, antimony, chrysotile asbestos, chromium (including ferrochromium) manganese (and manganese ferroalloys), the platinum-group metals, rutile and vanadium. In preparing the list, the working group assumed that the USSR and Eastern Europe were not reliable and secure suppliers.

The State Department study provides a sketch of each listed mineral with data on their functions and uses, substitutes, US production, current import sources, and supply availability outside southern Africa. Appendix B is taken directly from the State Department study.[17] There remains, however, considerable room for disagreement about the inclusion of certain items on this list, as will be explained in the next chapter on vulnerability. The important point, however, is that a consensus was achieved among the policy-making agencies involved. Although it is impossible to be entirely scientific in defining strategic minerals and reasonable people can disagree, the concern here is that when one overstates the degree of vulnerability, one is, in effect, propagating the notion that the world's largest, strongest, and most innovative economy could easily be brought to its knees by the loss of one or two strategic minerals. The record shows that this did not happen in World Wars I or II, the Korean war, or Vietnam, even though technology, transportation, information transfer, and industrial competitiveness were in their relative infancies. Indeed, the concerted efforts of some nations—through trade sanctions—to destroy the economies or bring down the governments of much smaller nations have failed in nearly every instance. Thus, the likelihood that the United States with its $4 trillion economy would be unable to survive the loss of certain supplies of strategic minerals approaches zero. Indeed, in this context, according to one unnamed administration official:

> If the whole continent of Africa, from Libya in the north to Cape Town in the south, sank beneath the waves, life would go on here just the same.[18]

On the "List of Ten" are four minerals—chromium, cobalt, manganese, and the platinum group—for which there is general agreement that their unavailability could indeed have a major impact on essential civilian and defense industries.[19] These four have few or no good substitutes, are essential to the production of important weapons or key industrial processes, and are located primarily in countries of questionable supply reliability—southern Africa and the USSR. A fifth mineral, vanadium, is also sometimes grouped in this category for essentially the same reasons.

The combined production of the USSR and South Africa, as a percent of the world's total, for four of these five minerals (cobalt being the exception) ranges from a low of 51 percent for manganese to a high of nearly 94 percent for the platinum-group metals (appendix A, table A-7). A similar calculation of reserves shows a low of 77 percent for manganese and a high of 98 percent for the platinum group. For cobalt, Zaire and Zambia combined account for some 70 percent of world reserves. However, counter to alarmists' views, a high level of supply concentration is only one factor affecting any vulnerability analysis and does not necessarily create a high degree of risk to the importing country. Many other factors need to be incorporated into the analysis and each mineral needs to be examined individually for the unique characteristics of its markets.

The Comprehensive Anti-Apartheid Act of 1986 also required an analysis of US dependence on South African imports for all materials in the government stockpile (appendix A, table A-8). The salient point is how sharply dependency percentages decline when the focus is on any one supplying country—even one as critically important as South Africa. Taking simple averages, total import dependence for the above four minerals averages 90 percent; for South Africa alone the percentage falls to 42 percent.

Analysis of other minerals in the study showed South African import dependencies of 28 percent for industrial diamonds, 29 percent for vanadium, and 24 percent for rutile, the primary ore for titanium metal. For most of the remaining materials in the stockpile, South African import dependencies were small or negligible.

Although South Africa is a key source for many strategic minerals, most of the cobalt imported by the United States is not produced there but in Zaire and Zambia. However, Pretoria's almost total control over the major transport networks in the southern Africa region raises valid concerns about Western access to cobalt supplies. Estimates of the percentage of cobalt which is shipped out of South African ports range from 50 to 90 percent. Intensive efforts to break the transportation stranglehold that South Africa exercises over its neighboring black states are being led by Prime Minister Mugabe of Zimbabwe. Of six routes to the sea that could by-pass South African rail and port facilities, only two are open—the Tanzam railway to Dar es Salaam in Tanzania and the Voi Nationale railroad and barge system to Matadi in the Congo—and these two routes are slow, inefficient, congested, and lacking in capacity. The other routes—through Mozambique to Nacala, Beira, and Maputo on the Indian Ocean and through Angola to Lobito on the Atlantic Ocean—have been closed for years due to internal guerrilla warfare. A regional transportation organization—

the Southern African Development and Co-ordination Conference (SADCC)—has ambitious plans to reopen or modernize most of these links.[20] The European Community, Scandinavia, and the United States have contributed $850 million to these projects, but some $4 billion will be required to finance the total reconstruction effort.

Among the first priorities of the SADCC is the reopening of the 400-mile rail, road, and oil pipeline connecting Zimbabwe's capital, Harare, with the Mozambican port of Beira.[21] Currently, about 90 percent of Zimbabwe's total exports and imports transit South Africa. There have been some unconfirmed reports that Pretoria could shut down Zimbabwe's entire mining economy within 30 days—including all ferrochromium exports. Aid officials, some members of Congress, and the State Department are reluctant, however, to put money into a project where security cannot be maintained. Even with 16,000 Zimbabwean troops in Mozambique, Maputo cannot secure the route against insurgency attacks by the anti-Marxist guerrillas. In addition to US political opposition against aid for Mozambique, an avowed Marxist government, budgetary pressures likely will limit US aid to levels well below those reportedly needed to carry out the project.[22] Whereas a fiscal 1987 bill before Congress requested aid of about $134 million over six years, the Reagan administration offered only $36 million in 1987 and $57 million in 1988—with none of the funds ear-marked for the Beira-corridor project. Given the political, economic, and security obstacles facing the frontline states, they will do well to modernize the Tanzam route to Dar es Salaam any time soon. While this project will alleviate some of the transportation pressures, it is unlikely that Zambia, Zaire, and Zimbabwe ever will be able to completely by-pass South Africa in shipping out their minerals, leaving these exports subject to some form of potential interdiction by Pretoria.

Critical Military Needs

The second part of the 1979 stockpiling definition emphasizes criticality of use. For defense-related end items, there is little ambiguity in this definition. Certain military items cannot be made without using strategic minerals. Among the best examples is the turbofan or jet engine used in the F-15 and F-16 fighter aircraft. Figure 6 shows the input requirements for the Pratt & Whitney engine used in these aircraft as of 1983. Each engine requires the procurement of nearly 5,500 pounds of titanium, more than 1,500 pounds of chromium, nearly 900 pounds of cobalt, and several other critical metals.[23] Columbium is another important jet engine metal for which the United States is 100 percent import dependent. Highlighting some other important uses of strategic minerals, chrysotile asbestos is needed for rocket and missile construction and vanadium for titanium alloys used in aircraft bodies. Armor-piercing shells generally cannot be made without tungsten. Military electronics and avionics make important use of platinum. Beryllium, one of the lightest metals, is used in missiles, aircraft brake disks and airframes, satellite and space vehicles, and inertial navigation systems for missiles and aircraft. Superalloys, which have a wide range of military uses because of their resistance to corrosion at very high temperatures, can contain as much as 65 percent cobalt and 25 percent chromium.

In most cases there are few acceptable substitutes for these strategic metals, although materials scientists are constantly investigating alternatives. The Department of Defense, for its part, has little incentive to adopt alternative materials—unless there appears to be an acute threat of a shortage or supply interruption for a given material in use because the development cycle for getting a new material into a piece of military hardware is lengthy, costly, and subject to risks. The decision to go with cobalt-based superalloys, for example, was made in the early fifties and only now are attempts being made to design cobalt out of these superalloys or at least to reduce their cobalt content. Perhaps the greatest obstacle to substitution in military end uses is that there can be no significant degradation in performance; i.e., there is little or no room for compromise in this area.

Despite these obstacles, research and testing continue—with an occasional breakthrough on the horizon. One particularly fruitful area—the development of cobalt-free oxide dispersion strengthened (ODS) superalloys, which are nickel and iron-based—could extend the operational life of jet engine combustor linings by up

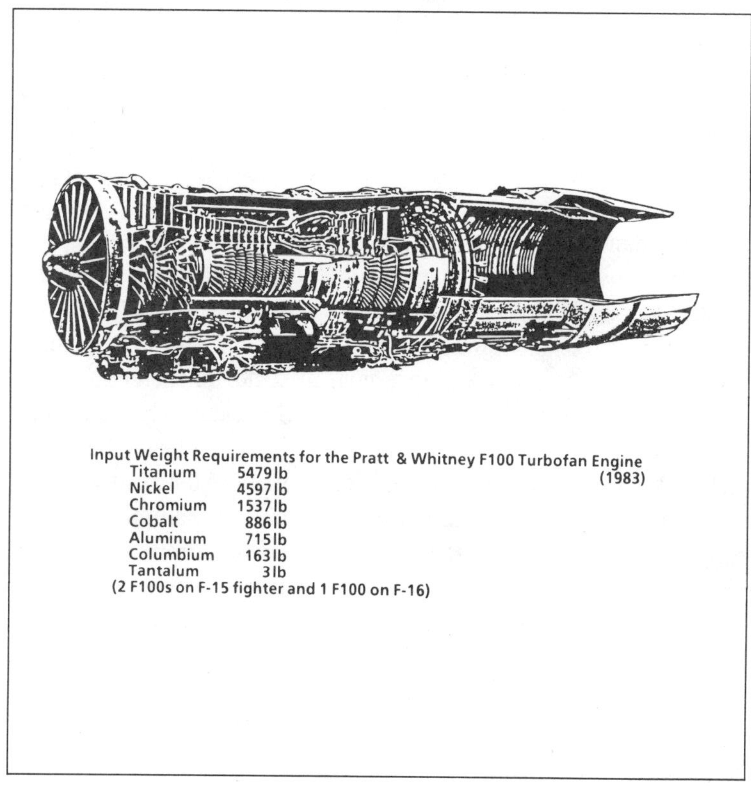

Figure 6. Strategic Metals in the F100 Engines.

to four times. Alloys of ODS materials also are generally superior at higher operating temperatures, but their inferiority at intermediate temperatures continues to be an obstacle to their use in jet turbine blades. However, one ODS alloy has been used for years in high-pressure turbine valves of military aircraft, and more widespread military applications of ODS technology can be expected.[24]

The greatest degree of ambiguity of "criticality of use" is found in "essential civilian" applications. No clear-cut definition exists within the US Government of essential civilian needs; perhaps they are best defined by what they are not. Clearly, private sector consumer-oriented industries generally are not considered to be essential. There is general agreement that the iron and steel and ferroalloy industries, the chemical and petrochemical industries, the machine tool industry, and the nuclear power industries are essential. Even these industries, however, produce both essential and non-essential products. There are a number of gray areas as well, such as the automotive industry. Clearly the ability to produce trucks—both military and nonmilitary—is critical. On the other hand, even the production of automobiles and the support industries involved could be considered essential, based on domestic economic criteria.

Using a broad definition, it would seem that any material shortage that brought a major industrial sector of the economy to a virtual standstill ought to be considered critical. This definition would encompass not only medium and heavy industry but high-tech industries, such as computers, as well. Along these lines, DOD and stockpile planners are beginning to examine the criticality of nontraditional materials such as gallium, germanium, and the rare earths. To that end, President Reagan's revised stockpile program provided, for the first time, for the purchase of germanium for the National Defense Stockpile.[25] On the other hand, there currently is no provision for the purchase of some two dozen other exotic materials necessary for the development of President Reagan's "Star Wars" system.

Looking Ahead

Having set the stage in this chapter by examining import dependence at some length and attempting to define what is meant by strategic minerals, the next chapter pursues in more detail a number of issues that have been only touched upon so far—such as vulnerability, substitution, and the stockpile. Chapter 3-"The Public Good and Strategic Minerals Policy"—will tie these issues to the policy process. In chapter 4, the author's perspective will be set aside in order to present the viewpoints of several key policymakers, as well as selected views in private industry. Industry experts have often expressed the opinion that their expertise has gone largely untapped in the making of strategic minerals policy and that their viewpoints remain unheard or ignored. In the final chapter, an attempt will be made to tie together the author's own analysis with the opinions and analysis of selected government and private experts and to recommend a new approach to the strategic minerals dependence problem.

2

A Key Distinction: Dependency Versus Vulnerability

THE COMMONLY HELD PERCEPTION THAT THE UNITED STATES IS in an especially vulnerable position because it depends so highly on imports for its strategic metals needs is pervasive and charged with emotion. In a July 1986 speech on South Africa, President Reagan stated, "Southern Africa and South Africa are the repository of many vital minerals . . . for which the West has no other secure source of supply." Similarly, President Botha of South Africa also engaged in a bit of hyperbole at times with his implied threats to cut off strategic minerals exports to the West, stating, "if South Africa were to withhold its chrome exports, one million Americans would lose their jobs" and "the motor industry in Europe would be brought to a standstill."[1] Neither President Reagan's statement nor Botha's claims are correct although both have some elements of truth.

The pervasiveness of the vulnerability issue extends far and wide. Even the American Legion has a position on the issue, adopted in September 1984 at its 66th National Convention in Salt Lake City:

> Resolved, that we support the US Government's policy of constructive engagement with South Africa, promoting peaceful evolution away from apartheid, and urge all Americans to recognize the important US strategic interest in South Africa stemming from its

> mineral wealth, key geographic location, developed economy and staunch anti-communist policies.[2]

This chapter will attempt to cut through the emotional and political rhetoric, examining in detail the difference between dependency and vulnerability in order to identify the areas where the risks of import dependence do, indeed, create problems for the policy-maker.

Distinguishing Vulnerability from Dependency

Import dependence—or more properly, as noted in chapter 1, the import deficit—is a necessary but not sufficient condition for vulnerability. To be vulnerable means to be open to attack or damage, implying a condition of weakness. Even this definition is not especially useful. Inherent to it are value judgments: attacks by whom and under what conditions and with what impact? It is this ambiguity that gives rise to a disparity of viewpoints and a lack of national consensus about whether the United States has a serious problem and, perhaps more important, what needs to be done about it.

In order to eliminate as much of this ambiguity as possible, experts have identified various measurable criteria that can be analytically applied to the problem of vulnerability. Each expert, however, has his own set of criteria, the formulations of which incorporate a weighted system of factors subjectively considered to be the most important elements in the overall calculus. For example, one expert study concluded that 16 criteria should be used to determine import vulnerability:

—Number and sources of supply and their location;
—Total US consumption from overseas sources of supply;
—Degree of importance to the US economy;
—Ease with which industry can substitute for the material;
—Ease with which the material can be recovered or recycled;
—Rate of increase in consumption;
—Need for the material by the military for national defense;
—Dollar amount used by the United States;
—Importance to the economies of the nation's allies;
—Extent of worldwide competition for dwindling supplies;

—Length of time required to expand sources of supply;
—Sudden shock impact of interruption of supply;
—Time required for substitution;
—Probability of an interruption, and length of time likely to be involved;
—Extent to which the material is used as a catalyst in a chemical process, i.e., the leverage factor;
—Political and economic aspects of supply.[3]

Even this grocery list of criteria fails to capture all of the elements that need to be considered in a sound vulnerability analysis. Missing from the list are such variables as the availability and adequacy of the National Defense Stockpile; the resource sharing agreement that exists between the United States and Canada; the vulnerability of US allies; US legal authority under the Defense Production Act to acquire and allocate scarce civilian supplies of materials under emergency conditions; costs of recycling and developing substitute materials; research initiatives on substitute materials by the Bureau of Mines, other federal agencies, and the private sector; the adequacy of the defense industrial base; and several others.

Perhaps the greatest flaw in most analyses is a failure to distinguish supply interruptions in peacetime from demand shortfalls coupled with supply interruptions under wartime conditions and the full mobilization of the US industrial base. Supply-side factors are the most relevant in the former case, while in the latter both supply and demand factors need to be taken into account. Indeed, a construct for vulnerability analysis that distinguishes supply and demand factors simplifies the analysis and provides a more orderly approach to understanding the problem. The key points of the discussion to follow include

—Supply concentration
—Reliability of supply
—Other sources of supply
—Recycling
—Effect of higher prices
—Technical conservation
—Substitution

—Wartime vulnerability considerations
—Industrial mobilization
—Vulnerability analysis: four case studies
—Pseudo-strategic minerals
—A new breed of strategic materials
—Other vulnerability assessments

Supply-Side Factors

Supply concentration. As suggested in chapter 1, vulnerability is proportional to the degree of concentration of foreign supply sources. The greater the degree of concentration, the greater the likely impact of a supply interruption. Among the Big Four strategic minerals, platinum-group metal production is the most highly concentrated, with South Africa and the USSR accounting for 95 percent of global production, and manganese production the least concentrated—the additive output of seven countries is necessary to reach the 95 percent level. Among other strategic materials with very high supply concentration levels are andalusite (sole producer—South Africa), spinning grade chrysotile asbestos (sole producer—Zimbabwe), and columbium (Brazil and Canada produce 99.7 percent of world output). Among other strategic minerals with relatively low degrees of production concentration are antimony (18 major suppliers), cobalt (8 major suppliers), tungsten (16 major suppliers), and bauxite (14 major suppliers). Thus, for those minerals with a low level of supply concentration, the loss of supplies from a single country not only would have minimal initial impact, but other suppliers would soon take up the slack by expanding their own output.

Reliability of supply. Vulnerability is also a function of the likelihood and expected duration of any potential supply disruption. In order to make such a determination one needs to assess the past and expected future reliability of the key supplying countries. Much has been written on supply reliability. Indeed, most major corporations have experts in political risk analysis—an area where political judgments tend to supersede data analysis. Despite a lack of consensus, a reasonable argument suggests that, in general, one ought to consider US friends and allies as the most reliable suppliers, LDCs as somewhat less reliable, and Communist country suppliers the least reliable. Another consideration—the length of supply lines—suggests that supplies from producers in the Western Hemisphere ought to be considered more reliable than supplies from Asia and Africa, particularly under a wartime scenario, since long supply lines are subject to a greater risk of interdiction. Depending on the scenario then, supplies from Australia and Thailand could be considered both reliable and unreliable. A third consideration attempts to make the distinction between sporadic or unplanned interruptions and the potential for government-initiated or planned disruptions in supply.[4]

The historical record shows a mixed picture of reliability. Since World War II, supply problems have pestered metals users. Interruptions, however, have been brief and mitigated by market adjustments, inventory drawdowns, conservation, and the substitution of other materials. There is no conclusive evidence indicating that increased import dependence has been accompanied by greater supply problems. Indeed, improvements in transportation, communications, and marketing, coupled with greater resource availability, suggest the opposite may have occurred.

Nonetheless, disruptions of supply occasionally do occur. They can stem from a number of factors—including natural disasters, strikes, and economic and political decisions. The effect of natural disasters on worldwide mineral supplies, however, has never been great. When such disasters occur, remedial action is usually undertaken quickly to restore production and exports to their former levels. On the other hand, mining activity in the *developed countries* has a long history of labor turmoil. Well-organized and militant unions in developed countries have frequently interrupted production with lengthy strikes. For example, the nickel industry in Canada was subjected to a six-month strike in 1980. In Australia, the mineral resources boom is being restrained by labor relations problems.

The supply record of the Less Developed Countries generally has been good—even though labor in the LDC mining industries generally is less disciplined than in the developed countries. Although wildcat strikes do occur, they usually are settled quickly. Thus, metals users have not found LDC unreliability to be a serious problem. Because trade is so important to their economies, Less Developed Countries go to great lengths to assure that mineral exports are not interrupted. Both state-owned and private mining companies frequently carry large inventories as a safeguard against strike-induced interruptions. Such inventories are used during slack periods to ensure that the orders of the best customers are met. According to industry estimates, for example, Zaire has about 20,000 tons of cobalt stored in Belgium. This amounts to about one year's consumption for the entire non-Communist world. Material such as this in the processing and marketing pipelines could keep supplies flowing to the West for a lengthy period even if mine production were interrupted.

The record of the *Communist countries* as minerals suppliers contrasts somewhat with that of the LDCs. Although these countries generally honor their contracts, they are in other ways unpredictable suppliers. Both the USSR and China have a history of bouncing in and out of the metals markets, especially in titanium (the USSR) and tungsten (China). Neither country has revealed much about its minerals policies, its future production and export plans, or the timing of marketing decisions. Moreover, because their mineral exports are state controlled, they can be quickly and effectively cut off. At the beginning of the Korean war, the USSR imposed an embargo on shipments of chrome to the United States. During the Vietnam conflict, on the other hand, Soviet chrome shipments to the United States increased. Similarly, following the imposition of trade sanctions by the United States against the USSR in January 1980, the Soviets continued to make deliveries of strategic metals under prior contracts and, in fact, solicited additional transactions. The strategic planner must, nonetheless, regard the Communist countries as unreliable sources of strategic minerals. As for marketing behavior, the Soviets generally follow business-like practices. No one has uncovered any evidence that the Soviets have tried to deprive the United States or the West of strategic metals in the market places. The Soviets, nevertheless, trade shrewdly and are sensitive to situations in which they can garner higher prices. In the 1970s they took advantage of the world chrome shortage brought about by UN sanctions against Rhodesia to triple the export price of Soviet chrome ore.[5] Other Western exporters tripled their prices as well. Similarly, Moscow has used its dominant role in the platinum-group metals trade to maintain high prices by limiting the volume of its exports.

The risk of government-sponsored actions designed to restrict accessibility to minerals supplies also is of concern to the West, but the historical record shows few successes. The ease with which the OPEC oil cartel was able to restrict world petroleum supplies and extract concessionary profits at the expense of the West has raised the specter that nonfuel minerals producers could do likewise. The West's fears appear to be unfounded. Nonfuel mineral cartels have been abysmal failures. The association of major copper exporting countries tried and failed to restrict output and maintain prices. The International Bauxite Association, formed by Jamaica in 1974 and followed by other Caribbean bauxite-producing countries, attempted to garner windfall profits by imposing stiff export taxes on local producers. Their ultimate failure was assured by Australia's refusal to participate in the association. Indeed, new investments immediately moved from the Caribbean nations to other countries while Caribbean enterprises stagnated. The demise of the producer-controlled International Tin Council in 1986—which became financially insolvent when it could no longer afford to buy up excess world tin supplies in an attempt to keep prices artificially high—eliminated the last cartel-like organization in the nonfuel minerals industry. Even the century-old DeBeers diamond cartel is foundering, as the world's three largest producers—Australia, Zaire, and the USSR—have begun to market more production through their own trade associations.

While such failures do not preclude future cartel attempts, such risks appear small as, in fact, the National Commission on Supplies and Shortages concluded,

> Minerals embargoes deliberately directed at the United States are only remotely conceivable. Embargoes directed against all importing countries are out of the question. Attempts to create minerals cartels may be made, but producers must reckon that the conditions for success are not usually present and that the cost to them of failure may be high.[6]

The conditions for success referred to by the commission are five-fold. First, the number of exporting countries must be small in order to get agreement on prices and market share. Second, demand in the long run must be inelastic; i.e., sharply higher prices must not move consumers to buy less. Third, supply by nonmembers must be inelastic; i.e., higher prices must not stimulate additional production by nonmembers. Fourth, the bonds among cartel members must be strong enough to prevent members from cheating or withdrawing from the cartel.[7] This latter consideration has contributed to the declining influence of OPEC in recent years. Finally, there must be no major exporter outside the cartel—as was the case when Australia refused to join the IBA. These conditions are unlikely to be found in any major mineral commodity market. Thus, those who talk of the threat to US security from minerals cartels appear to be riding the emotional waves of the OPEC upheaval.

Embargoes are another method of government-imposed trade restrictions that are perceived as potentially affecting US mineral supply availability. The notion that by exerting economic pressure, one can compel foreign governments to change their policies has always been a popular one, but successes have been few. A notable example was the 1980 US embargo on grain sales to the USSR in protest over Moscow's invasion of Afghanistan. In the end Soviet troops remained in Afghanistan while the United States lost a sizable share of the grain market to its major competitors—Australia, Canada, and Argentina. In minerals, the 1965 UN embargo on chromium purchases from Rhodesia appeared to have little effect on Rhodesia and a perverse effect on the United States, which had been a major importer of Rhodesian chrome. According to news accounts in 1971, even during the US sanctions Rhodesian mines were running at full output, and the ore made its way to markets in the developed West. France, Japan, Switzerland, and probably others reportedly purchased Rhodesian chrome at discount prices, disguising it as ore from South Africa and Mozambique. The United States, meanwhile, had to pay premium prices for Turkish and Soviet ore. Moreover, the Soviets were suspected of reselling Rhodesian chrome as their own. Thus, US dependence on Soviet supplies doubled between 1965 and 1970.[8]

In sum, embargoes are largely unsuccessful because of (a) a lack of unity by the world community against the targeted country; (b) cheating; (c) opportunities for middlemen to turn a profit; and (d) the fungibility of minerals, metals, and other commodities. A fifth factor—the rise of the multi-national corporation—has also facilitated the circumvention of most embargoes. As noted earlier, cheating was commonplace during the Rhodesian chrome embargo. The important role filled by brokers in normal commodity trade also facilitates the shipment of embargoed commodities to illegal destinations. Brokers or middlemen simply fill buy and sell orders from their customers at a standard service charge and generally have no legal or moral obligation to investigate either the source or destination of the commodities to be delivered; i.e., they are "disinterested" third parties. Any country or firm intent on making an illegal sale needs merely to arrange an indirect transaction through third parties who are not participants in the embargo in question. East European countries are particularly anxious to step in under such circumstances because of their need to generate scarce hard currency earnings.

The minting of the American Eagle gold coins provides an elegant example of the complexity of the market place. To set the stage, the Gold Bullion Act of 1985 required that the American Eagle coins be minted entirely from "US newly-mined gold."[9] The Comprehensive Anti-Apartheid Act bans the importation into the United States of the South African gold Krugerrand coins. However, at least 44 percent of US gold mining capacity is foreign-owned. Of this total, Canadian firms own 31 precent and a single Canadian firm—Anglo American Corporation—owns 10 percent. A US firm—Engelhard—mints more than 80 percent of the gold blanks for the Eagle coin. Engelhard has refineries in London where it is still legal to import South African bullion. In addition, 30 percent of Engelhard is owned by Minorco—a wholly-owned, Bermuda-based subsidiary of Anglo American. The normal refinery industry practice is to combine gold from all sources. It would be prohibitively expensive for refiners to segregate their gold supplies or to identify sources for each gold bar refined. Therefore, it is highly probable that the so-called "all-US" Eagle coin contains gold not only from South Africa but from the USSR as well—the world's second largest exporter of gold bullion after South Africa. Thus, Congress' well-intentioned embargo, imposed for political purposes, is being emasculated by the economics of the market place. Indeed, it is unlikely that a fool-proof embargo could ever be devised, at least in the area of strategic mineral commodities. Even a more stringent US embargo—one that might include, for example, bans on the purchase of South African chromium, manganese, or platinum—would likely be doomed from the start. It would likely have little impact on US supply availability or US strategic mineral vulnerability because South Africa would continue to sell to countries not participating in the ban. This metal would eventually find its way to the United States.

Other sources of supply. The strategic mineral vulnerability of the United States during a peacetime supply cutoff from one or more sources also would be mitigated on the supply side by the market's reaction to the higher prices that would result. The adjustment process would proceed roughly in the following manner. Initially, private inventories would be drawn down. These inventories in many instances are considerable. While rational companies attempt to minimize operating inventories because of cost, they tend to err on the conservative side because a halt in production generally means a halt in revenues and an inability to recover even their fixed costs. If companies are able to anticipate pending supply problems, higher-than-normal inventories are carried. For example, anticipated shortages of rhodium—used in catalytic converters—stimulated world automotive companies to stockpile this material. In general, private inventories could be expected to carry most strategic metals users through shortage situations lasting anywhere from 3 to 12 months.

Examination of the trend data of US stock levels for selected minerals shows a high degree of variance (appendix A, table A-10). At one end, stocks of platinum-group metals equivalent to more than 15 months of annual consumption and for vanadium, 8 months. Chromium ore, cobalt, and manganese stocks range from four to six months. For antimony, tungsten, and titanium sponge, stock levels are only one month of consumption. Looking at trend levels, the data show a very mixed picture. In some cases—titanium sponge and manganese—1987 stock levels have dropped to roughly half of 1985 levels. In five instances, current stocks have changed little since 1985, and in the case of platinum, the stock level has more than doubled. The key point illustrated by the data in the table is that stock levels reflect supply, demand, cost, and price levels, i.e., the economics of the market for each individual mineral. Indeed, based on these data, there appears to be no reflection of increased concern by producers or consumers about any increasing supply uncertainties because of geopolitical factors—despite the increase in the political and economic instability in South Africa since approximately 1983.

The relative price trends for the big five strategic minerals—chromium, cobalt, manganese, platinum, and vanadium—tend to support this conclusion of unconcern. With the exception of platinum, year-end 1986 price levels were well below levels in 1981-82. The price-level spike for platinum shown in figure 7 is explained by the sizable entrance of speculators into the market. Industry estimates of the speculative demand for platinum showed a 300 percent rise worldwide during 1985-86, equivalent to 15 percent of estimated total Free World demand for 1986. Investor-speculators were of course looking to "make a killing" in the platinum market, based on fears of a loss of South African platinum supplies. Many, however, probably were dismayed when platinum prices—which peaked at roughly $600 per troy ounce—began to fall to more normal levels in late 1986. This price correction reflected two factors: (a) a disruption in South African platinum supplies failed to materialize and (b) plans were well underway by South African platinum producers to expand production at currently producing mines and to begin production at entirely new platinum ore deposits.

The adjustment process under a supply shortfall would find other producers beginning to expand operating capacity. Currently, world *excess mining capacity* is near an all time high, the result of overinvestment in the 1970s and recession and unusually slow economic recovery in the 1980s. This condition is likely to persist until the economies of Western Europe and the LDCs begin to grow again at something approaching historical rates. Examination of world production and capacity figures shows the magnitude of

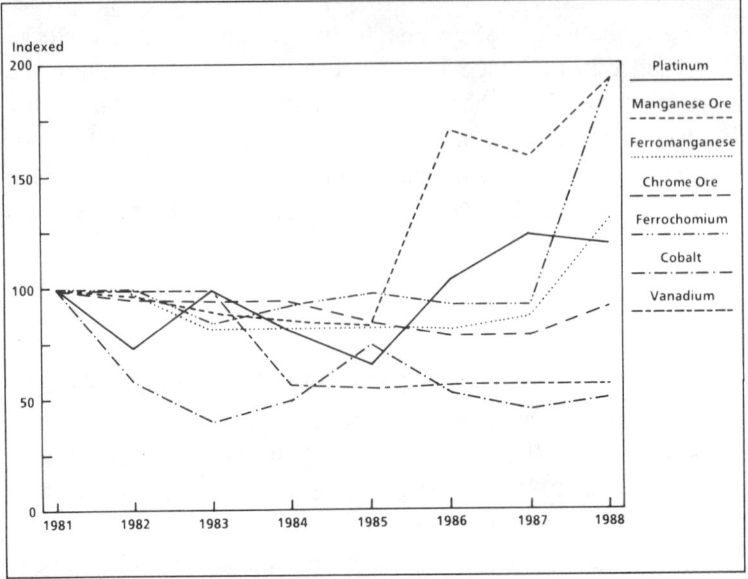

Figure 7. Key Metals: Relative Price Trends.

global overcapacity in the mining sector (appendix A, tables A-11 through A-16). In manganese ore production, average world operating capacity is only 70 percent. Excluding the USSR, the rate is even less at 63 percent. Of greater significance is the fact that Free World excess capacity for manganese is 2.5 times current South African production. As a result, the complete loss of South African production could be offset by stepping up production in the other producing countries. In assessing vulnerability, one needs to do this type of analysis on a mineral by mineral basis. At the other extreme, however, total Free World excess production capacity in platinum group metals is only 15 percent of current South African production. Thus, the loss of South African supplies could be only partially offset by production elsewhere.

If a supply disruption were an extended one, new capacity could also be brought on stream.

Again, one needs to proceed on a case by case basis to determine by how much the new capacity could offset a potential loss of supplies. In the case of a "greenfield" deepshaft mining operation, lead times of 7 to 10 years are not uncommon. Clearly, such new capacity would be of little help in the case of a sudden supply interruption. However, some mines such as surface mining and placer operations could be brought on stream in two to five years and could be of considerable value in expanding new supplies of strategic minerals in the event of an extended disruption in supplies, especially if warning indicators preceded the disruption.

Although the United States does not now mine chromium, cobalt, manganese, or the platinum-group metals in any significant quantities, it has done so in the past with the assistance of government subsidies under Title IV of the Defense Production Act. Bureau of Mines estimates indicate US production potential for these minerals (appendix A, table A-17). The estimates indicate that the United States has the potential to produce anywhere from 7 to 60 percent of its needs from domestic resources, with the best potential in cobalt and in the platinum-group metals, and that (with the exception of manganese) initial production could be brought on stream within two years. The US vulnerability to a platinum or palladium disruption is being alleviated somewhat by the start-up of production in 1987 at the Stillwater Mine Complex in Montana. According to estimates by the producers, as long as current platinum prices remain at or above $400 to $450 per troy ounce, mining operations there will be able to produce at a profit without government subsidies.

The potential for expansion outside the United States is considerably brighter. Major nonproducing platinum deposits exist in Canada, Zimbabwe, Brazil, Colombia, and Australia (appendix A, table A-18). The largest untapped cobalt resources are found in Canada and Uganda with less-rich resources existing in Peru and New Guinea. For manganese, Australia, Gabon, Brazil, and India could significantly expand output, given higher world prices. Rich, untapped chromium resources are to be found in India, New Caledonia, and the Philippines, while significant expansion of existing production could take place in Brazil, Finland, Greece, India, Madagascar, and Turkey. Reliable data on potential new production from these overseas sources are scarce, but some order-of-magnitude estimates have been made by the Office of Technology Assessment. Well over 12 million pounds per year of cobalt could be mined from 10 new sources, of which half could begin production within three years. This amount would be equivalent to about one-third of current global cobalt production. For manganese, a new mine in Brazil alone will add 500,000 tons of capacity within two to three years, with potential capacity of one million tons under the right market conditions. This mine alone could eventually offset the loss of 25 percent of South African manganese production.

Among the greatest constraints facing Brazil and other LDCs in developing such projects, however, is a lack of adequate infrastructure—storage, rail, shipping, and port capacity. The speed with which LDC strategic mineral resources could be developed would depend not only on the urgency to do so and on market conditions but also on their ability to generate adequate investment to develop these resources. Without resolution of the LDC debt problem and a more conducive global investment climate, many potential projects are unlikely to come to fruition. Unless the developed West begins to take steps now—through joint development and financing programs—few of these projects are likely to be capable of offsetting potential supply disruptions in the near or interim term.

Recycling. Western strategic mineral vulnerability could also be reduced by increased recycling. One should think of scrap material as an above-ground storehouse of strategic minerals, most of which are now routinely discarded. Although recycling is currently an expensive process because of the small scale on which it is now carried out, the potential exists to add significantly to available world supplies. The 1978-79 cobalt crisis provides a good example of what can be done in this area rather quickly, given the need to do so. Stimulated by a ten-fold rise in the producer price of cobalt as the result of a perceived shortage situation, alloy producers, parts fabricators, and turbine manufacturers began to segregate fabricated scrap by alloy type so that it could be remelted and reused.[10] The exact amount of savings provided by recycling during this period is not known but has been estimated at 10 to 25 percent of consumption.

The use of recycled material in the United States is relatively low (appendix A, table A-19). Identification and sorting of scrap is time consuming and expensive. Recycling of complex materials such as superalloys is technically complex and requires detailed knowledge of scrap constituents.[11] Nickel- and cobalt-based superalloy scrap may contain as many as 10 separate alloying elements as well as troublesome contaminants that need to be removed during processing. In stainless steel scrap, trace quantities of phosphorous can adversely affect workability, formability, and ductility. In tool steels, small amounts of titanium can be harmful. In many cases expensive x-ray spectrometers must be used to identify constituent elements in more complex materials. However, recent advances in user-friendly computer sorting techniques and the development of inexpensive mobile instruments have reduced sorting and identification costs and allowed greater use of technicians and unskilled operators. Once identified, normal processing techniques can generally be used to extract the desired elements. The major obstacle to the commercialization of advanced superalloy reclamation is still operating cost. While a small-scale commercial facility designed to process 100 pounds of superalloy scrap for contained cobalt is estimated to cost only $5 million, annual operating costs of $1.3 million would be unprofitable at current cobalt prices.

The potential for the recovery of chromium and platinum from secondary scrap is enormous, but each of these metals faces its own unique obstacles to greater recovery. It is estimated, for example, that some 24,000 tons of chromium are lost each year in recycling stainless steel scrap for carbon steelmaking. (Stainless steel contains from 12 to 30 percent chromium depending on the grade.) Another 38,000 tons is lost in unrecovered old stainless steel scrap. Only about 30-40 percent of the stainless steel contained in scrapped automobiles is recovered. Each automobile contains nearly 5 pounds of chromium with the untapped potential for recovery from this single source alone totaling some 5,000 to 6,000 tons per year. Another 17,400 tons of chromium is lost each year in metallurgical wastes and 3,000 tons in chemical wastes.

As long as ferrochromium remains cheap—less than 40 cents per pound—and plentiful, this potential will remain unrealized. A sustained disruption in foreign chromium supplies could raise prices by several-fold, however, and stimulate the recycling industry to recover much of this lost material.

The largest untapped source of secondary platinum is found in scrapped catalytic converters. Based on a 50 to 60 percent recovery rate, catalytic converter recycling could add 400,000 to 500,000 troy ounces annually to PGM supplies by 1995, compared with an expected new vehicle demand for 1.4 million troy ounces.[12] The industrial refining capacity is already in place—major PGM refining facilities are operated by Johnson Matthey and Gemini Industries—but institutional factors as well as the uncertainty over future platinum prices are the major bottlenecks. Collection is the primary problem. Many cars are simply abandoned; others are simply baled or shredded; and some are exported. As a result 20 to 30 percent of all cars are never dismantled. Of those reaching dismantling yards, many do not have their converters removed because of low scrap converter prices. It is likely, however, that a large number of scrap dealers are simply stockpiling converters until prices rise and that a major disruption in overseas platinum supplies would bring these converters into the recycling market.

For all strategic minerals, the surplus-market situation since 1979 has inhibited the development of the recycling industry. Under a return to tight market conditions, appreciable increases in recycling can be expected as collectors and processors respond to increased prices. Moreover, such conditions would generate greatly increased volume within the recycling industry. This would create economies of scale, bringing recycling costs down sharply and generating profits not now available. A number of opportunities are available for the recovery of three important strategic minerals as well as surmounting technical, economic, and institutional barriers in the recycling industry (appendix A, table A-20).

How Demand Affects Vulnerability

Effect of high prices. The US vulnerability to supply disruptions would also be mitigated by the effect of ensuing higher prices on consumption. The more serious the cutoff, the steeper the subsequent price rise would be and the greater the decline in overall demand. The impact of this *economic conservation* would depend on the degree of demand-price elasticity for the mineral affected. All of us are familiar with the impact of higher food prices on our consumption patterns. When steak becomes too dear, we consume less of it and eat more chicken and fish. The same holds true for strategic minerals users except that there are far fewer substitutes available with which to switch. Even within the strategic mineral class, however, the number of substitution alternatives varies considerably. For manganese, which is price inelastic, there are no acceptable alternatives in steelmaking. Thus, higher manganese prices probably would have little effect on manganese demand in the next three to five years. Over the longer term, however, technical conservation factors would result in less manganese use per ton of steel produced. Toward the other extreme lie platinum and cobalt, which have relatively more substitution possibilities. Thus, higher prices would have a greater dampening effect on the overall demand for platinum and cobalt in many of their applications.

Data on demand elasticities for the strategic minerals are sparse and imprecise, but the 1978 disruption in the cobalt market as a result of the invasion of Zaire's cobalt mining region by Katanganese guerrilla forces provides a useful paradigm. The resulting rise in cobalt prices from $6.85 per pound to $47.50 per pound on the spot market resulted by 1980 in a 19 percent drop in cobalt consumption below what it would have been without the price rise.[13] According to a Charles Rivers study, cobalt use in magnets fell by 50 percent in three years with four-fifths of the decline due to higher prices and one-fifth due to a weaker US economy. Although it is difficult to predict the effect of disruption-caused price hikes on the demand for the other strategic minerals, in most cases it would be considerable.

Among the most important adjustments during a supply disruption is the fact that available supplies of strategic minerals would be allocated by the market to their highest valued end use. Such a phenomenon would assure that strategic military demand for these materials would by and large be met, with the shortfall taken out of non-essential uses. As a rule of thumb, peacetime military demand accounts for only about 10 percent of total demand (see figure 8). For example, a shortage of chromium would be unlikely to greatly affect its use in superalloys or essential stainless steel products. Thus, jet engine production or corrosion-resistant tubing used in the petroleum refining and chemical industries would suffer little. Instead, the use of stainless steel for kitchen sinks and decorative trim on automobiles and household appliances would decline. Whether essential nonmilitary uses might be impinged would, of course, depend on the seriousness of the overall supply shortfall and the capability of other sources to offset the shortage.

Technical conservation. In the event of a sustained supply interruption, industry also would take steps to use less of the disrupted material per end-use item—referred to as *technical conservation*. Current usage patterns for strategic minerals optimize technical performance while minimizing the cost of production. A sharp and sustained rise in minerals prices would stimulate industry to change production processes in the direction of a compromise between performance and increased cost. Because strategic minerals are cheap and abundant, they are in many cases "overused." Indeed, much material is lost or wasted in the fabrication process. A few examples serve to illustrate these points. Of nearly 8.2 million pounds of cobalt used in making superalloys in 1980, only 45 percent was contained in final parts; the remainder was lost in the production process. In jet engine production fabrication, losses of 87 percent are common and 95 percent not unheard of. The use of newly developed near net-shape processing techniques—precision casting, advanced forging, and powder metallurgy—will ultimately reduce these losses by a considerable amount. Recent advances in casting technology alone could reduce losses to as little as 5 to 10 percent. Pratt and Whitney has patented a new isothermal forging process that could reduce the buy-to-fly ratio from to 8:1 to 4:1 for high temperature superalloy jet engine parts.[14] Using new powder metallurgy techniques, gears and camshafts—parts that normally require considerable machining—can be manufactured to near net-shapes, thereby reducing the generation of scrap material. The combined use of powder metallurgy

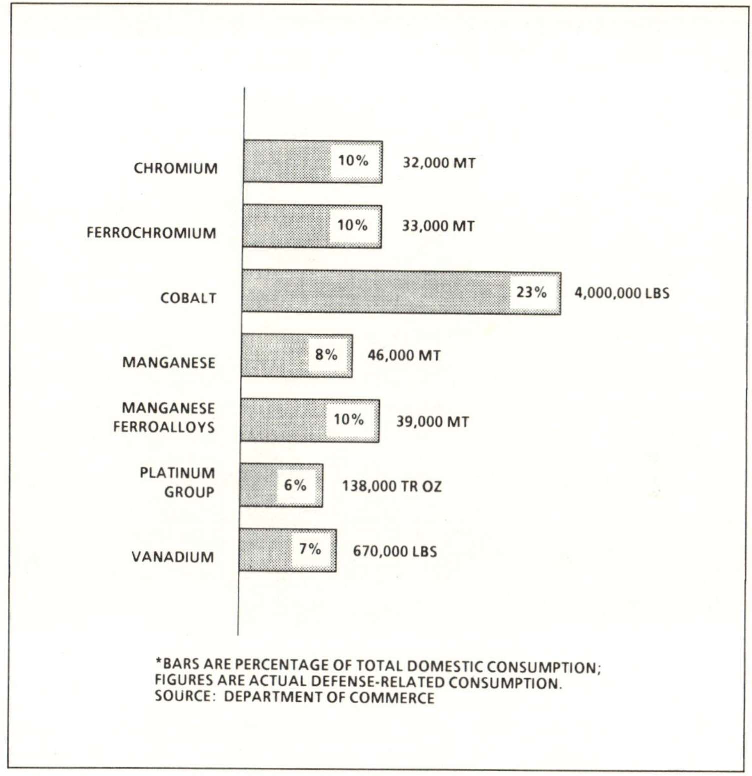

Figure 8. US Defense-Related Consumption (1983–84 average)*

and isothermal forging has permitted the manufacture of jet engine turbine blades from a new nickel-based alloy, reducing the need for chromium and especially cobalt. As with any major new industrial production process, start-up costs are high. Thus, while these technologies are now feasible, their rapid commercialization will depend on economic and strategic incentives to put them into place.

In the area of manganese usage where substitutes are virtually non-existent, new steelmaking processes could significantly reduce the need for this material over the next decade. The amount of manganese contained in a ton of steel depends on the steel type and ranges from 13 pounds per ton for ordinary carbon steel to 34 pounds per ton for stainless steel.[15] The average for the industry in the United States, given the product mix in 1982, was 13.8 pounds per ton. The production process, however, currently requires inputs of more than 35 pounds of manganese per ton of finished product. About 60 percent of this amount is lost in the form of slag, dust, and waste. New steelmaking technologies are expected eventually to result in a 12 percent reduction in the average manganese content of steel—from 13.8 to 12.2 pounds per ton—and a 42 percent reduction in consumption losses—from 21.8 to 12.6 pounds per ton. As a result of these trends, overall manganese requirements per ton of steel produced are expected to decline from 35.6 pounds to 24.8 by the year 2000—a 30 percent reduction. Given these trends, the Bureau of Mines forecasts an average annual growth rate in manganese consumption of only 1.9 percent for the United States and 1.4 percent for total world demand.[16] Current world manganese reserves are expected to be more than adequate to satisfy this demand.[17]

Substitution. Materials substitution provides another offset to US import vulnerability. The degree to which it can mitigate supply shortfalls, however, varies greatly by mineral and is affected by a host of technical, economic, and strategic factors. Because of its complexity, a few general considerations are in order. In the ideal, a substitute should satisfy several conditions: cost effectiveness, availability, performance, and compatibility.[18]

Substitute materials that are not readily available in adequate quantities or are not compatible with existing plant and equipment without major modifications are never developed for obvious reasons. In terms of technical performance, the substitute material should be comparable to the material being replaced. The degree of performance degradation that is acceptable, however, depends on the criticality of its end use. For defense purposes, any loss in performance may prove unacceptable whereas there is considerably more technical leeway in consumer-type products. In essential industries such as petroleum refining, chemicals, and nuclear power, small performance losses may prove acceptable if there are advantages to be had in other areas, e.g., greater supply availability or reduced production costs. In commercial operations, a more-expensive substitute would not be introduced, *ceteris paribus*. In contrast, the cost of materials is generally not critical in military weapons systems because they constitute such a small fraction of the final delivered cost of a weapons system. For example, the total cost of cobalt contained in the F100 engine is approximatly $8,700—about .03 percent of the cost of the engine ($250,000) and .00035 percent of the $25 million cost of the complete F-15 aircraft. Were an acceptable substitute for cobalt to be developed costing three times as much, it would increase the cost of the engine by only 1 percent and the final cost of the aircraft by only one-tenth of a percent.

Another type of substitution—functional substitution—can do away with the need for a particular material entirely. For example, with the development of cheaper aluminum gutters, copper gutters became obsolete in all but the most expensive homes. A return to the use of ceramic sinks and the wider adaption of fiberglass and composites could eliminate much of the need for chromium-containing stainless steel in non-essential uses. Technological innovation—driven by the desire to improve performance—is often at the root of such functional substitute development. The development of the tantalum capacitor is one such example, a significant improvement over previous types. This particular development, however, created a new import dependency. The United States now relies on foreign tantalum supplies for 91 percent of its needs.

For the most part, the substitute development process is quite lengthy. Lead times of 5 to 10 years are not uncommon—even in nonstrategic applications. Among the many steps involved along the way are innovation, research, gestation, laboratory production, testing and evaluation, commercial-scale design and engineering, costing, certification, and acceptance. Although few of these steps can be omitted, the time necessary to carry them out can be telescoped considerably under emergency conditions. The results, however, may sometimes be less than optimum. During World War II, for example, the United States substituted boron steels because of a shortage of high-strength, chromium-nickel alloy steels but at a cost of many fatigue-related failures.[19]

Substitution embodying "on-the-shelf" technology can also be carried out quickly, given sufficient economic or strategic impetus to do so. In such instances, the design, engineering, development and test work has largely been carried out, but the economics of the market militate against commercial introduction. When these economics change, e.g., due to a sustained shortage situation or a steep rise in the price of the material normally used, the substitute material can quickly become commercialized. Such was the case after the "cobalt panic" of 1978 pushed cobalt prices to more than seven times their normal level. "Easy substitutions" were first made in the magnet industry, and the price shock set the long-range substitution gears into motion in the superalloy industry. Other examples of easy substitution are found in dentistry and in electrical circuits. In these areas, gold, silver, platinum, and palladium are "near-perfect" substitutes for one another. The predominance of their usage depends almost entirely on their relative price levels. In some instances, technical substitution possibilities are restricted by public acceptance. In jewelry, for example, Americans prefer gold over platinum whereas in Japan platinum jewelry is more popular.

Substitution potential. Data that quantify the potential savings in strategic minerals usage provided by substitution possibilities are limited. Nonetheless, immediately available substitutes could replace one-third of the *chromium* now used, according to one estimate.[20] If fully realized, substitution could result in a one-time savings of 160,000 short tons of chromium—equivalent to 60 percent of annual chromium imports from South Africa. As noted before, there currently are no economic or strategic reasons for industry to move in this direction. Chromium is inexpensive and readily available and it would take an emergency situation—a major supply disruption or wartime mobilization—to provide the industry with sufficient incentives to make major changes in chromium usage. The research and development groundwork is being laid, however, for commercialization of certain chromium substitutes in the event of a supply cutoff, a surge in demand, or higher chromium prices.

Overall US *cobalt* consumption could be reduced by an estimated 40 percent, according to another estimate.[21] This would result in a savings of six million pounds of cobalt—equivalent to about 70 percent of annual US cobalt imports from Zaire and Zambia. The total savings would be comprised of a 70 percent reduction in the use of cobalt in hardfacing applications through the use of powder metallurgy techniques, an additional 50 percent reduction in magnetic applications, and a 60 percent reduction in chemical applications.

The potential for reducing *manganese* usage by using substitutes is negligible. For the *platinum-group metals* most quantitative studies have focused only on recycling potential. Similarly, little data exist on the substitution potential for most other strategic minerals. Such studies just have not focused on them because of their relative economic unimportance or because the risk of supply interruptions is lower than for the big four. Two exceptions are columbium and tantalum. *Columbium* imports come primarily from Brazil (73 percent of total US imports) and Canada (13 percent of US imports). In times of emergency, US consumption of columbium could be reduced by an estimated 20 percent in the initial phases of a supply disruption.[22] Similarly, *tantalum* consumption could be reduced by 20 percent under the same emergency conditions.[23] The savings would likely result from a 10 to 20 percent decline in cutting tool applications, a 25 percent reduction in capacitor applications, and a 30 percent reduction in usage in various types of steels.

Specific substitutes. The common notion that strategic minerals have no adequate substitutes is largely a myth. Such sweeping generalization perpetuates the fear of vulnerability. The lack of substitutes holds true only in certain specific applications. (See appendix A, table A-21 for substitution possibilities for five major strategic minerals.) Even manganese, which has no adequate substitutes in steelmaking (which consumes more than 70 percent of its usage), has some substitutes in its minor application (in chemicals and batteries). The public interest is not well served by such distortions or misrepresentation of the facts. Nor is it served by those with the polar opposite view that the United States is at no risk despite its high degree of reliance on imports. Vulnerability analysis must take all legitimate factors into account—including those substitution possibilities that exist, no matter how small—

if sound strategic minerals policies are to be formulated and implemented.

Vulnerability Under Wartime Conditions

The discussion of US strategic minerals vulnerability up to this point has been based on supply disruptions under peacetime conditions. A wartime environment introduces a very different set of considerations into contingency planning vis-a-vis strategic minerals. Under this type of scenario, one would expect to experience not only significant supply disruptions but also a sharp and steep surge in the production of military materiel and weapons requiring considerably greater amounts of strategic minerals. The record shows that the US capability to wage war in the past has in some cases been impinged on by a lack of adequate supplies of strategic minerals—in large measure because of a national stockpile that was insufficient (a) to offset supply-line interdictions or (b) to bridge the gap between normal industrial output and the time necessary to gear up for the sharply higher output needed to support the war effort. Although the balance of power and the ultimate outcome of past conflicts in which the United States has engaged itself did not ultimately hinge on the inadequacy of strategic minerals availability, former President Eisenhower avowed "lack of an adequate stockpile of strategic and critical materials gravely impeded our military operations. We were therefore forced into costly and disruptive expansion programs. The nation was compelled to divert, at a most crucial time, scarce equipment and machining and manpower to obtain the necessary materials."[24] Others have argued that past US war efforts have been unnecessarily extended by critical materials shortages and that the losses of lives and property were larger than they need to have been.[25]

Past mobilizations. Mobilization efforts by the United States, particularly during the first two World Wars and the Korean conflict, could be characterized as "erratic" and based on the "trial and error" approach. According to one expert, "The United States entered each conflict ill-prepared for the industrial production increases that eventually would be demanded to meet military requirements."[26] The US industrial support effort during World War

I has been characterized as "too little, too late." Only 145 75mm US field artillery guns and 16 tanks were shipped overseas prior to the armistice. Of more than 1,700 new steel ships ordered, only 107 were completed. American soldiers were largely equipped with French and British weapons.[27]

In contrast, World War II support efforts were monumental. The US defense industries produced great numbers of aircraft, tanks, battleships, destroyers, submarines, aircraft carriers, trucks, and rifles and carbines.[28] From 1944 to 1945 more than 40 percent of total US output went for war purposes.[29]

The figures, however, belie the fact that as a result of piecemeal government planning and general disorganization, US industry required three to four years—after receiving serious warning—to reach peak munitions production. With regard to manpower, draft deferments had to be extended to domestic mining industry employees to operate earthmoving equipment and to drive ore-carrying trucks—skills that could have been used overseas to build airstrips and access roads at the fighting fronts. Similarly, the domestic mining industry had to be given priority claim on bulldozers, scrapers, trucks, shovels, rock drills, and other scarce equipment.[30] In addition, a large share of the merchant fleet had to be diverted from the war effort to transport bulk imported ores from Latin America, Africa, Australia, and the Indian subcontinent.

The Korean war mobilization effort was generally more successful because of the limited nature of the conflict and the passage of the Defense Production Act of 1950 and because defense-owned plants remaining from World War II were able to facilitate industrial expansion. As a result, production of military supplies and equipment reached seven times the pre-war effort, investment in plant and equipment increased by more than 50 percent, and the size of both the armed services and national materials stockpile doubled. Moreover, new weapons systems were developed at an incredible rate. For example, 7 new aircraft were in production by 1953, and 16 more were under development.[31] Lead times were still a problem, however. More than two years after the conflict began, over two-thirds of the aircraft, missiles, tanks, trucks, and ammunition on order had yet to be delivered.

Representative Items Procured by Armed Services During World War II

	Major Weapons Systems
10	battleships
27	aircraft carriers
110	escort carriers
45	cruisers
358	destroyers
504	destroyer escorts
211	submarines
310,000	aircraft
88,000	tanks
	Weapons
41,000	guns and howitzers
750,000	rocket launchers and mortars
2,680,000	machine guns
12,500,000	rifles and carbines
	Food Supplies
2,000,000	tons of potatoes (Army)
2,880,000	tons of flour (Army)
	Ammunitions
29,000,000	heavy artillery shells
100,000	naval shells, 16 inch
645,000,000	rounds of light gun and howitzer shells
105,000,000	rocket and mortar shells
40,000,000,000	rounds of small arms ammunition
	Transportation Equipment
46,706	motorized weapons carriages
806,073	trucks, 2½ ton
82,000	landing craft
7,500	railway locomotives
2,800	transportable road and highway bridges
	Communications Equipment
900,000	radios (Army)
	Clothing
270,000,000	pairs of trousers (Army and Navy)

Source: Merritt and Carter, *Mobilization and the National Defense*.

The lessons learned about surge production in earlier conflicts, however, seemed to have been forgotten during the Vietnam conflict. Several factors were responsible. The government neither declared a national emergency nor used its mobilization authorities. Regarding procurements, it operated on a business-as-usual basis, letting contracts on a competitive bid basis in the belief that the Vietnam conflict would be short-lived. Private industry responded accordingly. Industry reasoned that if the government did not consider the situation urgent, neither should they, preferring instead to fill commercial orders at higher profits and to maintain good relations with valued customers. Moreover, as the war became increasingly unpopular, US metals companies began to feel uneasy from a public relations standpoint about supporting the war effort by bidding on government contracts.[32] In the ultimate, the gains in industrial preparedness achieved during the previous two wars were largely undone during and subsequent to Vietnam. In 1976, for example, the Department of Defense Defense Science Board reported on a US defense industrial base in a state of deterioration.[33] The decline has accelerated since the 1976 report. According to data reported by the US Ferroalloy Association, the number of ferroalloy plants declined from 29 to only 17 by 1985, employment decreased from 8,500 to less than 4,000, and the number of ferroalloy products fell from nearly 2,400 to less than 1,200.[34]

Regarding shipping and supply line problems, the World War II experience was at best unsettling. Attacks on Allied transport ships by a relatively small number of primitive German submarines ranged from nettlesome to serious, especially in the waters off the Cape of Good Hope and in the Caribbean. According to the President's Materials Policy Commission,

> In the first 7 months of the last war, enemy action destroyed 22 percent of the fleet bringing aluminum ore to the United States, and oil and gasoline tanker sinkings averaged 3.5 percent per month of tonnage in use.[35]

Many believe that shipping losses in the next conventional war could be several times more serious, and the data would seem to support these arguments. Of 72 raw materials deemed vital to American defense and industry, 69 are imported by sea.[36] More important, 95 percent of these imports arrive on foreign flag vessels.[37] These data take on an increased degree of seriousness for contingency planning because of the significant decline in the size of the US merchant marine fleet (see figure 9). In 1950 US Government and private merchant ships totaled 3,500 vessels with a shipping capacity of 37.4 million tons. By 1985 the fleet had declined to only 788 vessels with a total capacity of 23.8 million tons. Moreover, of these 788 vessels, only 401 were listed as active.[38] In sharp contrast, the Soviet merchant ship fleet has mushroomed from only about 450 vessels to more than 2,500 vessels with a shipping capacity of nearly 24 million tons—a 1,400 percent increase since 1950.

If the United States had to rely on the US-flag fleet alone to carry imported ores in time of crisis, it would be extremely hard pressed to meet defense needs, much less civilian requirements.[39] Assertions also have been made that the demand for open ocean military escorts to protect merchant shipping will exceed the number of men-of-war available.[40] For its part, the USSR faces no such constraints and, with the largest submarine fleet in the world, could exact considerably greater US shipping losses than did the German U-boats in the event of a US-USSR conventional war. According to many within the US Government, these facts thus argue for an even larger National Defense Stockpile than was necessary during previous wars as an insurance policy against sizable shipping losses that could be expected to occur in a future conventional war.

Short war versus long war. In the post-Korean war period, contingency planning—as it relates to the size of the stockpile and mobilization of the defense industrial base—has hinged on a heated debate between "short war" and "long war" proponents. Although the debate has appeared in many versions, no one has expressed the issue any better than Timothy D. Gill in his book, *Industrial Preparedness*:

> The short-war theory contends that any conventional war with the Soviets will begin after little or no warning, be characterized by extremely high consumption and attrition rates, and end quickly, perhaps in weeks or months. The argument continues that if U.S. forces lack large stockpiles of war reserve material on hand to fight such a war, the Soviets would rapidly overcome them, thus forcing the U. S. into a decision to capitulate or escalate to nuclear weapons.

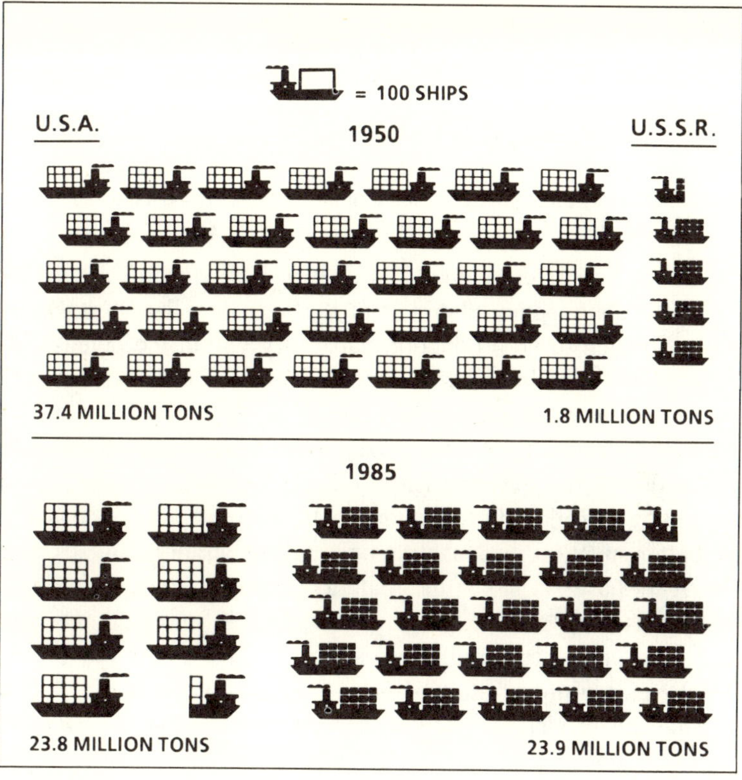

The size of the ships in the pictorial are proportional to the average carrying capacity for the years indicated.

Figure 9. Merchant Ships.

If the fight is to last such a short time, limited funds would be better spent on war reserve stockpiles to build a capability to counter and thereby deter the attack, rather than on preparations for industrial production. . . .

Those who believe that we should plan for a long war disagree with the short-war philosophy, arguing that it will become self-fulfilling [and could have an unfavorable outcome]. . . . Warfare involves too many unknown factors to be so certain of the duration of a conflict to dismiss the potential of the U.S. industrial base. Those who assume the next war will be short miss the opportunity to use one of the most important of U.S. assets. Furthermore, short-war theorists discount the usefulness of warning time, allowing

> industry to surge production . . . The short-war theorists also fail to account for the possibility of a conflict beginning in a region other than in Europe . . . and slowly escalating, again permitting production to surge. Finally, they ignore conflicts with adversaries other than the Soviets that would require significant conventional forces and time to successfully resolve. North Korea with a large modern army and aggressive tendencies comes to mind as a plausible candidate.[41]

To summarize: (a) the US-USSR nuclear standoff argues for long-war contingency planning in the form of a larger stockpile and a greater surge-production capability within the defense industrial complex in order to support it, or (b) the next war will be nuclear, over in short order, and there will not, therefore, be enough lead time to mobilize industrial, defense-related capacity or for that matter any necessity to do so, nor will there be a need for large stockpiles of critical materials, the funds for which would be better spent on nuclear weapons. Administration policies over the years have set stockpile goals to support a conventional war lasting as long as five years or as short as one year. Current stockpile goals are designed to support a three-year war effort. The Reagan administration planned for a long-war scenario, which is reflected in National Security Decision Directive (NSDD) 47 on mobilization, calling for a US capability to

—expand the size of US military forces from partial through full to total mobilization;
—deploy forces to theaters of operation, and sustain them in *protracted conflict*; and
—provide military assistance to civil authority, consistent with national defense priorities and applicable legal guidelines.

In detail current defense planning is designed to carry out the objectives of NSDD 47. The emphasis is on the need for increased surge production capabilities in order to overcome materials and production bottlenecks that could be reasonably expected during a national emergency, as well as the need to modernize the current National Defense Stockpile (see appendix C). President Reagan's stockpile modernization program—which would greatly reduce inventories of strategic minerals—generated considerable controversy within Congress and private industry. Opponents of the plan argued vehemently that sharply lower inventory levels called for by the new stockpile plan are not consistent with total force mobilization and are insufficient to bridge the gap between the onset of a national emergency and the time it would take to increase industrial capacity to a level sufficient to support an extended conventional war. Furthermore, some of these opponents suggest that the proposed stockpile cuts were made to generate revenues for the General Treasury Fund in order to reduce the budget deficit and were not, as the administration insists, the result of sound, impartial analysis of national defense related needs.[42]

Mobilizing the industrial base. The debate over stockpile adequacy and the adequacy of the defense industrial base is not likely to be resolved in the immediate future. Nonetheless, existing legal authorities concerning industrial mobilization considerably reduce US vulnerabilities for contingency planning under situations of a national emergency. Until recently, the Departments of Interior and Commerce and FEMA (Federal Emergency Management Agency) formed a triumvirate of agencies with considerable capability to mitigate the effects of industrial demand surges and supply cutoffs. As stated in the September 1988 *Strategic and Critical Materials Report to the Congress*,

> On 25 February 1988 the President designated the Secretary of Defense to be the National Defense Stockpile Manager. Previously, management and operations for the National Defense Stockpile . . . had been divided between the Federal Emergency Management Agency(FEMA) and the General Services Administration (GSA). Under Executive Order 12626, the President delegated to the Secretary of Defense all of his functions . . . except those provided in Section 7 (research and development activities), and Section 13 (imports of materials from designated sources). The President retains Sections 7 and 13; Section 8 was delegated to the Secretaries of the Interior and Agriculture. . . .
>
> The functions of the National Defense Stockpile Manager have been delegated to the Assistant Secretary of Defense (Production and Logistics) under the supervision of the Under Secretary for Acquisition. Certain operational activities relating to the National Defense Stockpile under Section 6. . . . have been further delegated to the Director, Defense Logistics Agency (DLA).

The Bureau of Mines and other agencies cooperate with DOD. A typical emergency situation would set into motion these contingency actions:

—Monitor and restrict exports that would drain scarce materials from the domestic economy.
—Under priority authority, fill rated (military) orders first; if necessary, allocate supplies among other claimants.
—Release stockpile materials on order of the President for purposes of national defense or in time of war declared by Congress.
—Initiate domestic and foreign supply expansion programs under Title III of the Defense Production Act.

The programs of the Defense Production Act were quite effective during the Korean war and the years following it—resulting in the initiation of US nickel mining and the creation of the US titanium industry and greatly increasing US production of aluminum, copper, and tungsten. In at least two instances—zirconium and helium—Bureau of Mines experts themselves initiated production to support the war effort because there was no US industry. In addition to direct production loans or loan guarantees under DPA, Congress has in the past, under special authorities, provided industry with more rapid depreciation benefits in order to stimulate private industry expansion. In other cases, where markets are not assured, Title III of the DPA provides authority for long-term government purchases.

The degree that the Defense Production Act can reduce US strategic mineral vulnerability depends in large measure on the amount of lead time available and the ability of military planners to identify their future needs well in advance of any emergency situation. As noted earlier, past failures in this area resulted in large military procurement orders going unfilled until the need for them had been overtaken by events. In the future, as military material needs involve more sophisticated high-tech materials such as gallium, germanium, and optical fibers, defense planning will become even more difficult. The traditional method of calculating defense needs as dollar percentages of civilian requirements using peacetime macro-economic models will no longer suffice. Rather than using the current disaggregation method, it is essential that military needs for these items be calculated on a unit basis, i.e., quantity per plane, missile, ship, tank, etc., and then aggregated if stockpile goals are to be accurately calculated.[43]

Vulnerability Analysis: Four Strategic Minerals

An accurate assessment of US vulnerability as a result of the overall high degree of reliance on imported strategic minerals cannot be accomplished based solely on generic arguments. One needs instead to look at the specific numbers for consumption, imports, stocks, foreign capacity, recycling, substitution, etc. Just as important, vulnerability analysis needs more specificity, i.e., vulnerability under what circumstances and situations. Thus, this next section attempts to put into concrete terms the general considerations discussed earlier in this chapter and in chapter 1 by looking at several specific strategic minerals on a case by case basis. The minerals analyzed here were selected because of their importance to the civilian and defense sectors of the US economy and because the risk of supply interruptions is relatively high. This is not to say that other minerals may not be equally critical to defense needs, but their uses are considerably more limited within the economy and their supply losses generally would have a lesser impact.

The following analyses assume that southern African supplies of these selected minerals are interrupted and that Zaire, Zambia, and Zimbabwe would have a great deal of difficulty exporting cobalt and chromium since they rely so heavily on South Africa's transportation network. Throughout the analyses, significant weight is given to the ability of market forces to adjust to these disruptions, as they did subsequent to the 1978 cobalt crisis in Zaire.

Manganese. A complete cutoff of South African manganese ore supplies would leave the Free World initially short by 1.34 million tons per year, and the United States short by only about 50 thousand tons. Gabon and Brazil are the other major US suppliers. The US producer and consumer stocks of nearly 600,000 tons could easily offset the loss of South African imports until other suppliers expanded their production. Within six months or less, excess Free World capacity totaling 1.4 million tons could be brought on stream, completely compensating for a permanent loss of South African production.

A cutoff in South African ferromanganese supplies would leave the world short by about 430,000 tons or 7 percent of total world output. The initial US shortfall would be about 150,000 tons or 37 percent of imports. Producer and consumer stocks would be sufficient to offset two-thirds of this loss or to maintain supplies for at least eight months. In addition, there is considerable world excess capacity available—over 30 countries produce manganese ferroalloys. According to 1984 data, the worldwide manganese ferroalloy industry was operating at only about 70 percent of capacity with world excess capacity of nearly 2.9 million tons, equal to nearly six times current South African production. Much of this capacity could be brought on stream rapidly if market conditions warranted such action.

An analysis of ferroalloy demand and supply under wartime mobilization conditions was completed in 1986 under a Department of Defense contract.[44] Even under some drastic assumptions about supply disruptions, the study concluded that there is sufficient worldwide manganese ore and ferromanganese processing capacity to meet military and essential civilian needs. For example, calculations show that the total US demand for steel in the event of a conventional war would annually reach 135 million tons, about equal to current US installed capacity. At this level, the demand for ferromanganese would be 844,000 tons. Even under the rigorous assumption that US ferroalloy capacity disappeared, and that all African and European supplies were totally disrupted, the United States would be able to meet its ferromanganese needs from other suppliers. In the event of supply line losses, which were not quantified, the study concluded that there was sufficient material in the national stockpile to meet any emergency. The government stockpile currently contains 2.24 million tons of metallurgical grade ore and 671,000 tons of high-carbon ferromanganese—equivalent to about one year's US consumption. During the next six years, under the stockpile upgrading program, an additional 472,000 tons of high-carbon ferromanganese will be added to the inventory.

Cobalt. South African cobalt production is insignificant compared to that of Zaire and Zambia, which produce roughly 80 percent of Free World supplies. An assessment of US cobalt vulnerability is complicated by transportation factors. As noted earlier, Zaire and Zambia rely heavily on South African rail and port facilities to export cobalt, but the percentage of exports reaching the West by this route is not known with much precision. Moreover, it is not known precisely what percentage of these exports could be diverted through export points in Tanzania and the Congo under emergency conditions. Third, significant amounts of cobalt could be air-freighted out of Zaire at an estimated cost of only $1 per pound.[45] Assuming that 40 percent of cobalt exports from Zaire and Zambia currently reach the West via non-South African routes and that half of the remaining 60 percent could be re-routed under extraordinary measures, the West would find itself short by about 4,100 tons annually. The United States, which is 60 percent dependent on supplies from Zaire and Zambia, imported an average of 4,600 tons of cobalt from these two countries during 1982-85 and under the above scenario potentially would face an annual prorated shortfall of 1,400 tons. However, private stocks in Belgium and in the United States could compensate for the loss of supplies from Zaire and Zambia for several years. Zaire currently maintains, at a minimum, a six-month supply of cobalt in warehouses in Belgium and reportedly plans to increase this amount to at least a one-year supply in order to firmly establish itself as a reliable supplier to the West. The US industry stocks, which stand at about 6,000 tons, would be sufficient to offset the above-calculated shortfall for four years, and the government stockpile of more than 26,000 tons contains the equivalent of more than 3 years annual US consumption or 5.5 years of average annual US imports from Zaire and Zambia. A major interruption in southern African cobalt supplies would, of course, create a panic mentality in the market and prices would zoom upward, eliciting new production by other producers. Excess capacity in Canada, the Philippines, Australia, and Finland—equal to 25 percent of current world output—would soon be brought on stream.[46]

If the supply interruption continued beyond one year, the development of new deposits would begin to take place. According to OTA estimates, well over 6,000 tons of cobalt per year could be mined from 10 new sources, half of which could be operational within three years.[47] The largest untapped resources are in Peru and New Guinea. Canada and Uganda have even larger untapped resources that would take longer to develop. Another cobalt crisis, such as the one postulated, would also stimulate a second round of conservation and substitution such as occurred in response to the perceived shortage of 1978-79. The United States could easily conserve about 20 percent of current needs and perhaps as much as 50 percent in three to four years according to one MIT estimate. With only half of potential new Free World capacity coming on stream and reductions in US demand due to price and conservation factors, the United States would be able to satisfy all of its cobalt needs, even with the permanent loss of production from Zaire and Zambia. If prices reached $25 a pound and remained there—about 3.5 times the current producer price set by Zaire— the United States could produce nearly 5,000 pounds of cobalt annually from resources located at the Blackbird Mine in Idaho, the Madison Mine in Missouri, and Gasquet Mountain, California.[48]

Chromium. A cutoff of South African chromium supplies plus the loss of an estimated 60 percent of chromium supplies from neighboring Zimbabwe would leave the West potentially short by about 1.1 million tons or about 50 percent of normally available Free World supplies.[49] Such a supply shock initially would create serious shortages especially if it came totally without warning. More likely, the cutoff would not be 100 percent immediately; significant quantities would be available in the shipping pipeline, and industry would have built up inventories—probably significantly—in anticipation of the disruption. Although supply bottlenecks would soon occur and prices would rise sharply, market adjustments would begin immediately to ameliorate the supply shortfall. Excess Free World capacity totaling 500,000 tons—equivalent to half the size of the disruption—would increase world supplies within a matter of months. A potential US shortfall of about 300,000 tons would be cut at least in half by producer and consumer inventories and by another 80 to 100 thousand tons, the prorated US market share of global new capacity brought on stream. The hypothetical US shortage of about 50,000 tons or 10 percent of normal consumption in the first year would have to be absorbed by a decline in consumption in non-essential uses as a result of higher prices. Defense-related consumption of about 50,000 tons would not be affected as this sector would simply outbid other users for available supplies.

In the event of a permanent cutoff lasting beyond 12 months, other market adjustments would mitigate the shortfall. Recycling and substitution could add the equivalent of about 50,000 tons to available US supplies. Over the long run, new capacity totaling some 800,000 tons could be added by new mine development in India, New Caledonia, and the Philippines and the expansion of existing mines in Brazil, Finland, Greece, India, Madagascar, and Turkey. These adjustment processes could result in reduced US demand of about 250,000 tons while the United States would have at its disposal non-African Free World supplies totaling some 300,000 tons. In the event that some of these demand and supply changes failed to materialize, the United States, in an emergency, would be able to draw on government stockpiles. The current chromium inventory is equal to more than two years of US consumption at current rates, three years at the reduced demand levels projected here, and more than a four-year equivalence of South African imports. In the ultimate, the United States could, according to Bureau of Mines estimates, produce 235,000 tons annually from domestic resources at double current price levels from deposits in Montana, California and Oregon.

Results of the DOD contract study on US *ferrochromium* processing capacity under mobilization conditions indicate that in the worst case—the absence of any US processing capacity and a simultaneous loss of capacity in South Africa, Zimbabwe, and Western Europe, there would be severe shortages of this metal. Under these extreme conditions, the United States would find itself short by 1.4 million tons of ferrochromium which could not be procured from abroad.[50] Current ferrochromium inventories in the stockpile could satisfy industry requirements for about 15 months. Chrome ore in the stockpile could meet US needs for an additional 15 months if it could be processed, but it would take 15 to 18 months or longer to build entirely new ferroalloy plants. Under the stockpile upgrading program, the United States plans to add about 50,000 tons per year of ferrochromium to the inventory over the next six years by converting existing stockpiled chrome ore.[51]

The pessimistic stringencies imposed under this worst case scenario flow from declines in the US ferroalloy industry. Between 1978 and 1985 domestic ferroalloy shipments fell from 1.6 million tons to 700,000 tons and employment declined by 50 percent. Similarly, processing capacity declined by over 40 percent. For ferrochromium, domestic shipments declined from 275,000 tons in 1978-79 to only 38,000 tons in 1984, and imports have taken over 90 percent of the US market. Nevertheless, barring further declines, US ferroalloy capacity stands at 1.2 million tons, and about 600,000 tons of standby capacity exists.

While the gloomy picture painted by the DOD contract study was probably intended to raise warning flags for DOD planners and the Congress, relying on worst case assumptions that are unrealistic and highly improbable does a disservice to the public interest and to the industry itself. The US ferroalloy industry is not likely to disappear completely, nor is the European industry. Indeed, US industry is undergoing a retrenchment designed to make it more competitive in the future. Although these are legitimate concerns within the industry, special interest group legislation likely to arise from such skewed analysis would tend to perpetuate the public perception that the US mining and metals processing industry seeks an undue share of the taxpayer's dollar. The only way to change this perception is through public debate and sound analysis and to show that a strong domestic industrial capacity is in the public's best interest from both a strategic and economic standpoint.

Platinum-group metals. A cutoff of South African platinum-group metals would reduce Free World supply availability by 60 percent or 3.6 million troy ounces. This total is comprised of 2.2 million ounces of platinum, 900,000 ounces of palladium and about 110,000 ounces of rhodium—based on 1986 Bureau of Mines data. For the United States, this would mean the loss of nearly 1.2 million ounces of platinum, 530,000 ounces of palladium, and nearly 100,000 ounces of rhodium. A total US shortfall from South Africa of 1.8 million troy ounces would equal more than 60 percent of US consumption in 1986. Excess Free World capacity of 340,000 ounces could offset only 10 percent of the total South African disruption while excess Soviet capacity, mostly in palladium, could offset another 300,000 ounces if it were made available to the West. In the event of a short-term disruption, US industry inventories of one million ounces would suffice for about only four months. An additional 500,000 to 600,000 ounces of platinum in dealer inventories in New York and London could provide the United States with an additional three-month cushion. In sum, assuming additional stock build-ups in anticipation of a South African cutoff, inventories and excess capacity likely would be able to weather a cutoff lasting something less than 12 months. Beyond that time period, potentially serious shortages could develop and a number of extraordinary adjustment factors would have to come into play.

As prices began to rise based on fears of shortages and the inevitable hoarding that would take place, significant substitution in the jewelry, dentistry, and electronics industries would take place. Demand for new platinum jewelry—currently taking about 30 percent of Free World supplies annually—would fall off sharply and old jewelry, which can be thought of as an above-ground mine of platinum, would begin to be melted down when the value of its contained platinum reached a high-enough level.[52] Electronics users would begin to substitute gold for platinum, and gold and silver would substitute in dental uses. Recycling of electronic scrap could provide 100,000 ounces of platinum per year and recycled catalytic converters as much as 500,000 ounces per year within five to seven years.[53] At some high price, platinum investors, speculators, and hoarders would begin disinvesting or dishoarding in order to take their profits, freeing at a minimum some 400,000 ounces. Domestically, production in the United States—which began on a limited scale in 1987 at the Stillwater, Montana, complex—is expected to double by 1990 and could provide an additional 150,000 ounces of palladium and 50,000 ounces of platinum annually.

Despite these developments, the United States would likely find itself short of platinum by several hundred thousand ounces under conditions of long-term cutoff in South African supplies.

Military needs—accounting for only about 6 percent of current consumption—likely would not be affected. Indeed, platinum and palladium needed for aircraft, missiles, satellites, electronics, and sensors (see figure 10) would continue to be available as defense contractors simply outbid other claimants on short supplies.

The greatest impact would occur in the US auto industry, which currently accounts for nearly 50 percent of annual consumption. The standard three-way catalytic converter requires .05 ounces of platinum, .02 ounces of palladium, and .005 ounces of rhodium. Assuming no technological change in these rigid ratios, a platinum deficit of 175,000 ounces, for example, would reduce the number of converters that could be built by some 3.5 million units—about one-third the number turned out in 1986, the auto industry's record production year. Similarly, each 10,000-ounce shortfall in rhodium availability could theoretically cut converter production by two million units. The auto industry even now is concerned about the adequacy of rhodium supplies in the absence of supply disruptions (for example, see the Ford Motor Company article, page 108). In 1986, the United States imported 98,000 ounces of rhodium from South Africa, 38,000 ounces from the United Kingdom, and 25,000 ounces from the Soviet Union. The total from the three countries was probably triple actual needs for converter production, the surplus demand being used to build inventories. In the case of an acute platinum or rhodium shortage that prevented the full production of catalytic converters by the auto industry, several options would exist:

—temporary suspension of current auto emissions standards; cars produced during the suspension period could then be retrofitted with converters once supplies became available.
—installation of programmed fuel metering systems for leaner engine performance; this would still allow average new car emissions to approximate 1979 levels.
—development of an alternate catalytic system using, for example, zirconium nitride; converters using this catalyst might last only 10,000 miles however.
—use of alternative engine technologies such as the lean-burn egine, which has very low emissions levels. Japan

SYSTEMS	STRATEGIC MINERALS IMPORTED FROM SOUTHERN AFRICA						
				Platinum-Group Metals			
	Chromium	Cobalt	Manganese	Iridium	Palladium	Platinum	Vanadium
Aircraft	X	X	X	X		X	X
Artillery	X	X	X				X
Ammunition	X	X	X				
Electronics	X	X			X	X	
Engines	X	X	X				X
Helicopters	X	X	X				X
Mines	X	X	X				X
Missiles	X	X	X	X	X	X	X
Satellites	X	X	X	X	X	X	X
Sensors	X	X		X	X	X	
Ships	X	X	X				X
Submarines	X	X	X				
Tanks	X	X	X				X

Figure 10. Defense Systems Requiring Strategic Minerals Supplied by Southern Africa.

and Western Europe are pursuing this technology aggressively.

—lower emissions standards; this would permit the substitution of more palladium for platinum in the typical converter, as is done in Japan.

None of these is a great option. In fact, pressure from environmentalists and the Environmental Protection Agency for considerably stricter emissions standards could very well obviate the use of all but two options unless the President or Congress stepped in to order a temporary suspension of current emissions standards under emergency authority. This could well happen because of the impact on the economy that a sharp reduction in automobile production would have. Absent such emergency measures, research on alternate emissions technologies would have to be stepped up considerably, putting to the test the innovative capability of US automotive research engineers. Talks with auto industry executives indicate that they are in the process of trying to come up with new designs that do not depend so heavily on PGMs, but they are not optimistic about achieving a breakthrough any time soon.

Pseudo-Strategic Minerals

The term *pseudo-strategic* is not found in any of the literature on strategic minerals. The term pseudo-strategic minerals as used here underscores the difference between legitimate vulnerabilities based on foreign import dependence and those cases where the risks of import dependence are small and manageable. In this exercise, four "so-called" strategic minerals were selected for analysis: antimony and industrial diamonds (found on the "List of Ten" referred to in chapter 1) and tantalum and tungsten, which show up periodically on lists of strategic minerals compiled by government analysts or by academicians. All four minerals are also contained in the inventories of the National Defense Stockpile. (See appendix A, table A-22 for summary statistics on the four.) If one were to examine the periodic table of the elements, the majority of them could be classified for an important military application, and the majority are not produced in significant quantities within the United States. In other words, if these were the only two important criteria, the list of strategic and critical materials would extend the length of one's arm and present for policy-makers contingency planning considerations of unmanageable proportions. Moreover, such a list would vastly overstate the degree of strategic vulnerability facing the United States. The following provides thumbnail sketches of the four materials, pinpointing the data that strongly suggest that these materials should not be considered worrisome for the United States:

Antimony. The strategic uses of antimony are minimal—for hardening small arms ordnance, for military vehicle batteries, and as flame retardants—probably totaling less than 7 to 8 percent of total annual consumption. Import reliance on South Africa is less than 14 percent, with China, Bolivia, and Mexico alone providing nearly 2.5 times the amount that comes from South Africa. More important, the stockpile contains the equivalent of 1.5 years of US annual imports, some of which is excess to the stockpile goal and is being sold off. Multiple accepted substitutes exist. In fact, with the development of a lead-tin-calcium alloy for battery applications, the poor prospects for electric vehicles, and the development of organic compounds for use as flame retardants, the primary problem facing the antimony industry is to develop new markets for this decreasingly important metal.

Industrial diamonds. The primary "strategic" use for natural stones is for hard-rock drilling. The natural stone market is under intense pressure from synthetics—prices of natural stones have fallen by more than 50 percent since 1982. Polycrystalline synthetics can now substitute for natural stones in most major applications. Australia, Botswana, and Zaire produce more than 80 percent of all natural industrial stones—which could be delivered by air freight in any emergency. The stockpile contains four years' worth of imports, and excesses are being sold. The United States, in fact, is a net exporter of industrial diamonds.

Tantalum. While the United States has negligible resources and depends on imports for about 90 percent of consumption, nearly 60 percent of US imports originate from supplying countries that are considered to be reliable—Thailand, Brazil, Australia, and Malaysia. Although strategic uses of tantalum are numerous, in electronics applications tantalum is being replaced rapidly by ceramics and aluminum-based materials. The most critical strategic uses are in superalloys, but the stockpile contains more than a three-year supply of US annual imports.

Tungsten. The world market is awash in tungsten with prices near a 20-year low. Although half of total usage is considered essential and nonsubstitutable, the stockpile contains nearly 17 years of US annual imports. Moreover, more than 10 Free World producers supply the market, and China is flooding the market with cheap tungsten. The use of coatings made from aluminum oxide and titanium carbide has extended the useful life of tungsten cutting tools, while cutting tools made from ceramics, polycrystalline diamonds, and titanium are also making inroads into this market. Although tungsten in superalloys is increasing, total usage is small and is replacing, in some cases, strategic metals that are less abundant. For some armor-piercing penetrators, tungsten is being replaced by depleted-uranium-based projectiles. As long as tungsten remains abundant and cheap, there will be little incentive to pursue additional substitution possibilities on a major scale or to explore for tungsten within the United States. Thus, import reliance will remain high.

Similar analyses could be performed on many other pseudo-strategic materials that would yield like conclusions. Summarizing the key points that often fail to be taken into account

—world resources outside of southern Africa are abundant;
—strategic-military usage, although often critical, is a small proportion of total US needs;
—imports are high and substitution research and domestic exploration are low because cheap world supplies are available;
—innovation in new products and technology processes are rarely factored into conventional analysis; and
—the National Defense Stockpile, in most cases, provides more than an adequate cushion against conceivable supply emergencies.

Unless these points are given adequate weight, the result is likely to be a skewing of recommendations—especially by alarmists—toward inefficient policies that exaggerate the degree of US vulnerability and frustrate rather than serve the public interest. Moreover, as the revelations of the Vietnam war so graphically demonstrated, once the public loses faith in the integrity of those in positions of public trust, the backlash of public opinion can do irreparable damage to the ability of policy-makers to deal with *legitimate* security threats against the United States.

A New Breed of Strategic Materials

The rapid pace of technological change in the West is creating a new breed of strategic materials. Although the need for traditional metallic minerals will never be completely eliminated, their relative importance is expected to decline with the advent of advanced ceramics, composites and micro-electronics necessary for the deployment of lasers, bubble memories, and more sophisticated weapons systems. Other materials will become strategic spin-offs of commercially derived products. With these changes, the role of the LDCs as suppliers will diminish in favor of Japan, Western Europe, and the Newly Industrialized Countries (NICs)—reshaping in the process the concept of foreign import reliance. In some cases, the concept of strategic materials may have to be broadened to include patented processes and technologies and the notion of a National Defense Stockpile perhaps altered to include the inventory of scientific data.

Along these lines, the Reagan administration allocated $30 million in FY 1987 for the acquisition of 30 metric tons of germanium—a by-product of zinc processing—and $24 million each for FY 1988, FY 1989, and FY 1990. Important new uses for germanium include high-data-rate optical communications systems, lasers, night-vision systems, and weapons guidance. Nascent studies are now underway to determine the criticality of other advanced materials and the possible need to begin stockpiling them. Other materials such as rhenium, gallium, and the rare earths could well emerge as additional stockpile candidates.[54] Although the United States has the world's largest rhenium reserves, the small size of the industry, declining industrial demand, and the high cost of its recovery—it is a by-product of a by-product—

make future domestic production uncertain. Its major strategic use is in jet engine superalloys. Gallium's major application is in bubble memories for computers and in gallium-arsenide-based integrated circuits. Gallium-arsenide-based circuits provide an order of magnitude improvement in circuit speed and are used almost entirely in military applications at the present time. Although the material is abundant, it is found almost entirely in spent tailings from the aluminum production process. Most production capacity is currently located in only a few plants in Western Europe, Japan, Czechoslovakia, and Hungary. A single US plant began production in early 1986. Much will depend on whether gallium-arsenide chips become the commercial standard and on who takes the lead in their development. A situation could develop wherein Japan becomes the world's dominant producer of such chips, with the United States relying on Japan to supply its military needs for such chips. The rare earths will be crucial to the development of the SDI system. Additional studies will be required to determine the sources and adequacy of supplies. Depending on the results of these studies, the rare earths could well be candidates for stockpile acquisition.

How Others Have Assessed Vulnerability

As seen in the four case studies, US vulnerabilities to supply cutoffs are actually quite limited in nature, with the major potential impacts likely to occur in chromium ferroalloys and the platinum-group metals. In no instances, based on the above analyses and realistic assumptions, would military needs be threatened. These conclusions are generally consistent with a study performed by the Interior Department. Although the methodology differs somewhat, the general consensus of the study is that US strategic minerals dependence—at least in a peacetime environment—is not a serious problem. The Department of Interior study, prepared at the request of Congress, concluded that

> for chromium, the President's modernized National Defense Stockpile (NDS), if fully utilized, could meet any shortfall originating from a three-year disruption in supplies from Zimbabwe and South Africa . . . that a manganese supply disruption from South Africa could be mitigated for nearly two years by domestic private stock drawdowns alone . . . that US producer, consumer, and dealer cobalt

> stocks could mitigate a supply disruption from southern Africa for 10 months while the proposed new National Defense Stockpile could do the same for an additional 20 months . . . and that a disruption of South African vanadium supplies would have minimal domestic impact because of a 21-month supply of private stocks.[55]

Although some private studies, many press articles, and even congressional testimony often cite specific military or civilian vulnerabilities, these vulnerabilities, as previously indicated, are often based on unrealistic assumptions or scenarios. It is true that *if* sufficient amounts of chromium *could not be obtained*, the US stainless steel and superalloy industries would indeed be crippled, seriously impinging on the production of missiles, ships, submarines, and fighter jets, as well as the chemical, electric power, and transportation industries. Similarly, *if* sufficient quantities of manganese and vanadium could not be obtained, steel production would decline; *if* cobalt were unavailable, the F100 engine, used in the F-15 and F-16, could not be built; *if* sufficient supplies of platinum could not be obtained, major technological changes would have to be made, at great cost, to the way gasoline is refined from petroleum, the way that auto emissions standards are met, the way certain fertilizers are produced, and the way that glass fibers for building materials and optical fibers for telecommunications systems are produced. But legitimate worst-case scenarios would not leave the United States and its allies in dire straits.

This is not to say, however, that there should not be constructive concern on the part of policy-makers and private industry about the situation. It is just such concerns, dating back to World War I, that have produced the myriad of laws, acts, and initiatives capable of dealing with the problem of vulnerability—uncovering, through greater awareness and research, the enormous potential of other offsetting factors (see appendix A, table A-22). Short-sighted policies, however, could easily undo the work of past administrations—for example, by letting the Defense Production Act lapse or reducing the size of the government stockpile below necessary levels. Similarly, sharp cuts in R&D funding, as a result of budget pressures, could reduce the ability of private industry to develop substitutes. Highly restrictive environmental laws could hamstring private development initiatives to mine strategic minerals from public lands or near national parks. Lack of planning by defense officials could result in serious new dependencies for foreign sources of high-tech materials with strategic uses. These issues, which are addressed in the succeeding chapters, demand that a coherent and consistent strategic minerals policy be developed, implemented, and maintained if the vulnerability risks of import dependence are to remain both low and manageable.

3

Strategic Minerals Policy and Public Good

ATTEMPTS TO DEFINE THE "PUBLIC GOOD" HAVE A LONG HISTORY, ranging at least as far back as the *Federalist* papers of Alexander Hamilton. The theoretical debate has continued since then among philosophers, political scientists, and students of public administration. As a practical matter, however, a discussion of strategic minerals policy must take into account not only national defense considerations but economic and environmental issues as well.[1] Because the formulation of a strategic minerals policy has the potential to affect so many other areas with a diversity of concerns—especially in the economic arena—any proposed policy is certain to be controversial. To be successful, any such policy most certainly has to balance these diverse concerns. Thus, this balancing act of vested interests requires the establishment of priorities. Usually when determinations of public policy are made, there are interests that are satisfied and those that are dissatisfied; often no one is completely satisfied. In any circumstance, a policy must be justified not only to those specific interests that are affected but to the public at large. Not the least of a policy-maker's concerns, a proposed policy must be accepted by the legislative branch. The justification of a policy to the public at large and to the legislature in particular requires that the proposed policy be shown to benefit the entire public and not only a part of it.

Indeed, any public action assumes that there is a public interest and that a policy must provide for the good of the whole. Therefore, the public servant must justify the proposal in terms of changing something for the better or preventing a change for the worse, justifications that are almost always controversial. But the controversy serves as the focus of public debate, and the requirement that such issues be debated enables the nation to examine the justifications that are being advanced.

Clearly, the public interest may or may not coincide with what the general public or a part of the public demands or believes to be worthwhile. The responsibility of policy-makers is to identify where the public interest lies and to convince the public at large that such a policy is in everyone's interest, for a public official must be not only responsible but responsive. In other words, the public interest cannot be served unless the policy-maker is able both to sort out those interests that best serve society at large and to convince others, whether in Congress or in the executive branch, that a proposed policy represents what is good for society at large and not merely the special interests of a few. The difficulty of satisfying these objectives is well documented in American political history.

Formulating Strategic Minerals Policy

For purposes of the ensuing discussion, the *public good* is defined here as "safeguarding the national security of the United States, its citizens, and its Allies." This definition is quite similar to the definition of *national defense*. Indeed, strategic minerals form a subset of the national defense in that they are the input materials for weapons systems, the defense industrial base, and essential civilian productive activities.

At first glance, it would seem that this definition is rather straightforward, but closer examination shows that a considerable degree of subjectivity is embodied within it. Among the several considerations involved in the definition is the question, How much security is enough security? Conversely, how much risk is the United States willing to take, given that absolute security is an unreachable objective? Put another way, what degree of vulnerability is acceptable and under what conditions? For example, the risks and vulnerabilities under normal peacetime conditions differ appreciably from wartime possibilities, with the number of "wartime scenarios" ranging widely from low-intensity conflicts—in which the United States may play only a peripheral role—to full-scale conventional war between the United States and the Soviet Union, or its surrogates. Moreover, such a confrontation could be brief or of long duration. These scenarios involve judgments by US policy-makers about the likelihood and seriousness of possible emergencies, i.e., the educated "best guesses" of experts—none of which can be established in any scientific or completely objective manner. By way of illustration, civil war in South Africa would almost certainly lead to a disruption of key mineral exports from that country, but the degree and duration of such an interruption has been the subject of considerable debate by experts in the Department of Defense, the intelligence community, and the Department of State.

At the other extreme, an all-out conventional East-West war would impinge on many strategic mineral supply considerations and be affected by the length of the conflict, theaters of operation, number and types of weapons to be used, attrition of men and materiel, protection of shipping lanes, and the ability of the economy to mobilize its domestic mineral resources and to surge industrial production. Given the number of variables involved, even sophisticated computer models are unable to generate meaningful results without a number of simplifying assumptions. Here too, judgments are involved in determining which assumptions are realistic. The Department of Defense, which attempts to determine material requirements, has been openly criticized by the General Accounting Office concerning both the methodology used and assumptions made in establishing its materials requirements.[2]

A final consideration, and perhaps the most important in practical terms, is cost; i.e., what price are we willing to pay to assure an "adequate degree of security." The size of the National Defense Stockpile—which can be thought of as an insurance policy—provides a concrete example of disagreement over the security-cost issue. For example, in 1937, the Navy Department was authorized to purchase $3.5 million worth of materials determined to be necessary in time of war.[3] Subsequently, annual stockpile authorizations expanded to $100 million in 1937, $2.1 billion in 1946, and $3 billion by 1951. During this period, however, funds appropriated by Congress for the National Defense Stockpile fell considerably short of authorized stockpile goals. By 1962 the stockpile contained only $2 billion in materials.

The current National Defense Stockpile, based on goals established in 1980, contains inventories valued at $9.1 billion—of which $2.0 billion is considered to be material in excess of individual material goals—leaving unfilled goals at $12.3 billion as of 31 March 1988 prices.[4] In sharp contrast, based on the results of a study conducted by the National Security Council (NSC) during 1984-85, President Reagan on 5 July 1985 recommended to Congress a stockpile of only $700 million plus a supplemental reserve of $6 billion.[5] The President also recommended the sale, over five years, of surplus inventories totaling $3.2 billion. Congressional and private industry opposition to this plan was considerable. In addition, various government studies have attempted to weigh the trade-offs between stockpiling materials and subsidizing their domestic production under the Defense Production Act in an attempt to determine the best approach for increasing the supply availability of certain specific materials. These studies note that each unit of annual production of these materials is able to offset the need for three units of the same material in the stockpile, based on the assumption of a three-year conventional war.

With the strategic minerals' public good competing with the public good in other areas, the policy-maker—whether in the executive branch or in Congress—faces a difficult balancing act—especially where large defense expenditures run headlong against the public's desire for clean air, protection of wilderness areas and wildlife, increased human services, and lower taxes. A stronger and more costly strategic minerals policy may be particularly difficult to sell to the American public for several reasons: it involves technically complex concepts such as superalloys and materials substitution, about which the public has little understanding; it relates to the defense sector, seen by a majority as an already bloated consumer of precious tax dollars in the absence of any real threat, and the ability of the United States and Soviet Union to destroy each other several times over; it involves the mining industry, which in the past has gained a reputation for despoiling the countryside, polluting the air, and receiving undue government assistance in the form of generous depletion allowances and other tax benefits while reaping large profits; and finally, it involves relatively few jobs now that high-technology and services have replaced mining and "smokestack" industries as major sources of national income. In addition, lobbyists for the environment have a much broader base of support than does the mining lobby. From a foreign policy perspective, US strategic mineral policy may be difficult to divorce from public hostility toward South Africa because of the emotional and moral objections to apartheid. Thus, the view of the public good by voters may vary considerably from the government view, and the policy-maker entrusted with national defense responsibilities may have to spend as much time educating and selling the public as actually formulating policy.

Even within the government itself, strategic minerals tend to take a back seat to other more pressing issues. Mexico and Brazil have considerable mineral resources, the development of which could diversify US import sources and increase reliability of supply. But more pressing issues tend to occupy the time and energy of policy-makers dealing with these countries. For Mexico, the issues of a rapidly declining economy as a result of falling oil prices and internal corruption, debt repayment, illegal immigration, and drug trafficking tend to occupy the administration full-time. Similarly, for Brazil the debt repayment problem, the viability of the current democratic regime, and agricultural trade issues tend to command the most attention. It is unlikely that any of these problems will be resolved soon. Therefore, unless US private industry takes the initiative to develop Mexican and Brazilian mineral resources, the United States is unlikely to benefit in any major way from their proximity. In contrast, other countries, such as Japan, are already staking their claims in the region through joint mineral ventures supported by government subsidies.[6] It would seem that bilateral talks should take into account

US strategic mineral needs as well, even to the extent of using a quid pro quo approach on certain issues—such as granting Brazil agricultural concessions—in order to make some gains in access to their strategic minerals. In the absense of putting strategic minerals on the front burner of discussions with the countries, one must question the seriousness with which the United States views the strategic minerals problems. In an integrated world economy, for the United States to limit prospective solutions to its own internal resources does not appear to serve the public good efficiently.

Last but not least among factors deciding the public good, policy-makers themselves are at odds over whether the United States is vulnerable to cutoffs of its supplies of strategic minerals. According to Dr. Charles L. Schultze, former Director of the Bureau of the Budget,

> Any modern industrial economy and particularly the United States, is incredibly quick to adapt to shortages of particular materials. Substitutes are quickly discovered, synthetics developed, and ways found to minimize the use of short-supply items. If all imports to the United States were cut off and every one of our overseas investments expropriated, the US economy would not collapse. Our living standards would suffer, but not by a large amount.[7]

Others take the view that a cessation of certain critical material imports could create economic disruptions more serious than any that might occur from a cutoff of petroleum supplies. During congressional testimony before the Senate Committee on Foreign Relations, one witness claimed that in the extreme, "the Nation's essential industries might be shut down within six months, and that continued lack of such supplies could cause a reversion of 40-50 years in the US standard of living and technology."[8]

Approaches to Risk Reduction

With such a disparity of views over whether a serious problem exists, one should not be surprised that no consensus exists on what, if anything, the United States ought to be doing in this area. Moreover, disagreement exists over how the public good is to be achieved, even among those who believe that the degree of risk is a major cause for concern. As a starting point, in this section the assumption is made that market forces alone cannot provide sufficient adjustments to completely compensate for serious disruptions in strategic mineral supplies and that some form of government action would be required to meet contingencies. Most policy approaches have tended to fall into one of five general categories:

—increased domestic production;
—larger stockpiles;
—resource allocation and sharing;
—greater diversity of overseas supplies; and
—technological solutions.

Indeed, all of the above have been tried in the past or are now being applied to some degree. Some "solutions" have been "one-shot" experiments; others have been based on existing standing authorities in the event of a materials supply emergency. Although it could be useful to go into exhaustive detail about each of the above policy approaches, a few succinct examples will serve to highlight past and present attempts to mitigate the risk of a strategic minerals supply disruption.

Increasing Domestic Production

Under Title III of the Defense Production Act of 1950, some $8.4 billion was spent during and after the Korean war to expand the production of strategic and critical mineral commodities through a system of price guarantees, purchase commitments, direct loans, loan guarantees, and tax benefits. The net cost to the government, however, after acquisition of these commodities from the private sector and the sale of unused materials was only $900 million. By 1956, DPA-supported production had resulted in a doubling of US aluminum capacity, an increase in US copper-mining capacity of 25 percent, initiation of US nickel mining, creation of a tantalum industry, and a quadrupling of US tungsten mining and processing. Title I of the same act, created specifically to deal with materials shortages during the Korean war, established a system of priorities and allocations that (a) made Defense Department orders the first claimant against the output of strategic and critical materials, (b) distributed the entire remaining supply according to a priority system developed by federal emergency management agencies, and (c) issued conservation and limitation orders to limit or forbid certain non-essential uses. Title I authorities also were used extensively during the Vietnam war. Similarly, tax incentives in the form of accelerated depreciation and investment tax credits were used in the early 1950s to stimulate major expansion of the US industrial base. Rapid depreciation alone—over a five-year period—added nearly 25 percent to US steel capacity. Tax laws continue to encourage mining expansion by granting higher depletion rates for minerals considered more strategic.[9]

In contrast, under peacetime conditions, the government generally has been content to let free market forces determine US mineral supplies. This passive approach to the problem worked well during the boom years of the 1950s when supplies of all commodities from oil to agricultural goods became increasingly tight and fears of world shortages stimulated massive private investment. However, since the early 1980s' recession and the failure of the Western economies to recover to previous high economic rates of growth, the US minerals industry has fallen on hard times. Much production and processing capacity has moved offshore as foreign competition has stiffened. From 1982 to 1986, for example, the value of US nonfuel minerals production—especially metallic minerals—virtually stagnated, contributing to a share decline in the US defense industrial base. Figures for 1987 and 1988 reflect rising prices and output.[10]

Although the steel and copper industries have been hardest hit, strategic minerals also have fared poorly. Nearly flat demand continued in 1986 for refractory metals, nickel and chromium-bearing stainless steels, alloy and specialty steels, chromium and tantalum for aerospace uses, and oil drilling equipment containing cobalt-bearing tungsten carbides. Until recently weak demand and high inventories generally have held down the price of nickel, chromium, tungsten, columbium, tantalum, and molybdenum—exacerbated by imports of low-priced metals. The single US nickel

Value of US Nonfuel Mineral Production

	Metals	*Industrial Minerals*	*Total*
1983	6,004	17,151	23,161
1984	5,621	17,612	23,233
1985	5,608	17,872	23,480
1986	5,817	17,642	23,459
1987	7,444	18,902	26,346
1988	10,428	20,032	30,460

Million current dollars; US Bureau of Mines.

mine was forced to close in mid-1986. Operating rates for ferroalloy producers averaged less than 50 percent. As a result of these trends, employment in the nonfuel mineral mining and processing industries has declined by 18 percent or about 600,000 jobs since 1981. As a point of contrast, spurred by a continuation of generous government subsidies, US grain production continues to reach record or near-record levels year after year and government-held grain inventories are now sufficient to satisfy total world import demand for two years.

The influx of cheap imports has resulted in a negative nonfuel mineral trade balance with the deficit for 1986 estimated at $15 billion.[11] Although inexpensive imports may mean cheaper prices for consumers, they contribute to greater unemployment and continued erosion of the US industrial base, a matter of considerable concern to policy-makers. Lobbyists have argued that the United States should take some form of action, such as imposing import quotas or tariffs to counter unfair foreign competition. The Reagan administration, however, argued that the technological inefficiency of domestic producers led to the state of affairs in the minerals and metals processing industries. The Reagan administration took this position in late 1985 when it refused to provide relief to US copper producers against cheap Chilean copper imports. Mining interests, on the other hand, counter that foreign producers are heavily subsidized by their governments, making it impossible for US producers to effectively compete on a worldwide basis.

The situation is further complicated by production policies in the Less Developed Countries; when prices fall, they tend to increase output in order to sustain the hard currency earnings needed to service their foreign debts and to maintain domestic political stability and employment—problems not likely to be resolved for many years. Failure to produce all-out would create for them an even more precarious economic situation, and the accompanying higher unemployment would increase the chances of domestic instability. Recent administrations have tended to give these foreign policy considerations a great deal of weight and have tended not to take protectionist actions against cheap imports for this reason.

Another approach to greater domestic production, especially for minerals in which the United States has so far identified only marginally economic resources, is to open up public lands to mineral exploration and development. This approach has been hindered by several factors: legislative-regulatory gridlock, technology problems, and investment disincentives. According to a 1979 survey, 40 percent of the nearly 825 million acres of federal land was formally closed to mineral development and another 10 percent is highly restricted in such use.[12] This acreage is reserved for wildlife protection, national parks and recreation, agriculture, energy development, and military and Indian uses. This "single-use" approach to federal land management has created a massive roadblock to domestic mineral exploration—despite section 2 of the Mining and Minerals Policy Act of 1970, which requires the government to "foster and encourage" the multiple use of public lands. Indeed, Congress, especially since 1976, has moved increasingly against the intent of this law through other legislative avenues restricting public land usage.[13] Major lobbying efforts by the National Wildlife Federation, regulatory restrictions by the Environmental Protection Agency, the efforts of pro-environment members of Congress, and judicial inaction on these issues have brought mineral development on public lands to a virtual standstill, despite the efforts of the Reagan administration, which returned several million acres of public lands to the multiple use category.

From a technical standpoint, even if federal lands were opened to exploration, many experts feel that the prospects for major discoveries are slim. Precambrian rocks are considered the most geologically favorable areas for the existence of chrome, cobalt, and platinum deposits, but such geologic formations are not extensive in the United States. Other geologists disagree with this majority opinion, however, pointing out that increased geologic knowledge, better satellite reconnaissance technology and fresh exploration concepts could find new deposits. It is believed, for example, that the technology level employed by mineral geologists today is about 20 years behind the level used in oil and gas exploration.[14]

This fact ties in closely with a third obstacle—a lack of reliable long-term economic incentives to explore for strategic minerals. The abundance of cheap foreign supplies and the expected continuation of low prices for the foreseeable future have removed most incentives to undertake risky—as well as costly—exploration programs. New mining ventures can cost $1 billion or more from exploration to development. By way of contrast, the total value of US imports for five major strategic minerals—chromium, cobalt, manganese, platinum group, and vanadium—is only about $5 billion. Thus it is not surprising that no group is actively exploring for strategic minerals in the United States today. Unless government resources in the form of funds or technology are applied to the problem, prospects for a near-term discovery breakthrough appear to be slim.

The above discussion does not, of course, pertain to already known sub-economic deposits of strategic minerals. Dozens of such deposits exist from Maine to California and Alaska, but these are uneconomic to develop at current or expected long-term price trends (appendix A, tables A-23 through A-26). Some mines, in fact, which have been in production in the past, have since been shut down. For example, the most recent US production of chromium occurred at the Stillwater Mine Complex in Montana during 1953-61. About 900,000 tons of ore were produced at a government-guaranteed price of nearly $35 per ton under the Defense Production Act. This price was more than four times the world price. Similarly, US cobalt production ceased in 1972. During

1948-62, some 14 million pounds of cobalt were produced domestically under stockpile purchase contracts and the DPA program. During World War II, manganese ores were produced in 20 states—satisfying 13 percent of US requirements—but all production ceased by 1970.[15] In the lone exception, the 1986 rise in platinum prices to well over $600 per troy ounce stimulated the private development of platinum production at the Stillwater Mine in Montana. According to projections, with a sustained market price of around $450 per troy ounce, the mine is expected to produce as much as 100,000 troy ounces of platinum and 340,000 troy ounces of palladium by 1990, which could satisfy about 15 and 30 percent, respectively, of US needs. Prices for 1987-89 have hovered in this range. Market prices would have to increase severalfold before production of the other strategic minerals would be economic. In all of the above cases, potential domestic production would still satisfy only a fraction of US needs, even with government assistance.

Stockpiling: A Study in Frustration

The subject of stockpiling has drawn more attention and generated more controversy than any other aspect of the strategic minerals issue. Certainly there has been more congressional involvement, policy recommendations, and legislation on stockpiling than for any other aspect of the strategic minerals problem. The concept of a National Defense Stockpile has its origins in materials shortages experienced in World War I. Although congressional hearings on the subject go back to 1880 and planning was begun by the Army General Staff in 1921, it was not until 1937 that the Department of the Navy was authorized to purchase $3.5 million worth of material deemed to be necessary in time of war.[16] The National Strategic Stockpile concept was not institutionalized, however, until passage of the Strategic and Critical Stockpiling Act of 1939, which also authorized the purchase of materials for essential civilian and defense-supporting industries, as well as wartime materials. The $100 million purchase program authorized for the period 1939-43 was interrupted by the outbreak of World War after only $70 million had been spent.

A greatly expanded Stockpile Act was passed in 1946, which empowered the government to assume control of all inventories of strategic and critical materials, as well as to promote domestic production of these items. A stockpile goal of $2.1 billion was established which, after allowing for stocks already in the inventory, left the stockpile short by $1.8 billion. The gap was to be closed by annual purchases of $360 million per year over five years, but because Congress annually appropriated considerably smaller sums, the stockpile was only 40 percent complete at the onset of the Korean war. Subsequently, within a period of only six months, nearly $3 billion additional was appropriated. The National Stockpile was also expanded as a result of two other stockpiles. Excess materials totaling some $2 billion, purchased under the Defense Production Act, were transferred to the National Stockpile by 1962. Similarly, strategic materials were acquired for the National Defense Stockpile by bartering surplus agricultural commodities under the authority of the Commodity Credit Corporation Act of 1949. It was not until the passage of the Strategic and Critical Materials Stock Piling Revision Act of 1979 that all government stockpiles of strategic materials were consolidated into one inventory.

Over the years US stockpile policy has suffered the slings and arrows both of Congress and the executive branch, with Presidents playing a particularly active role. Under the Eisenhower administration, the stockpile objective of sustaining a five-year war effort—established during World War II—was reduced to only three years, creating large surpluses for disposal. Additional surpluses were created in 1973 when the Nixon administration reduced this objective to only one year, subjecting itself to criticism that the reduction was made to generate large revenues designed to reduce the budget deficit—a deficit that was becoming a political liability. Manipulation of the stockpile for economic purposes was not confined to the Nixon administration, however. In 1965 President Johnson threatened to release 300,000 tons of stockpiled aluminum in order to force the aluminum industry to rescind a major price increase. Congress acted to halt all disposals in 1975 until a new stockpile study could be conducted. The subsequent inter-agency study—led by the National Security Council—reaffirmed the three-year objective and announced a major, long-range disposal and acquisitions program.

Until late 1986, fragmentation between policy-making offices and their operational components also existed within the stockpile program. Prior to 1973, the Office of Defense Mobilization was responsible for stockpile policy and the General Services Administration (GSA) for stockpile management operations. In 1973 both functions were posited with GSA, but the 1979 Stockpiling Act again split off the policy-making component, transferring it to the Federal Emergency Management Agency (FEMA). By early 1987 all stockpile functions were transferred to FEMA. Although this structure would appear to simplify matters greatly, annual stockpile transactions were determined by a steering committee chaired by FEMA but with representation from Agriculture, Commerce, Defense, State, Energy, Interior, Treasury, the Office of Management and Budget (OMB), GSA, and CIA. Each agency had an interest in the effect that a purchase or sale might have in those areas for which it has some responsibility. Thus, there is an attempt to factor in considerations such as national security, mobilization needs, international and domestic economic policy, the US industrial base, domestic resource development, and foreign policy into decisions to buy or dispose of strategic and critical materials. Although the goal is altruistic, the results have been less than satisfactory.

Congress, too, has tended to weigh in heavily when it perceives that a particular stockpile plan runs counter to interests that it represents. For example, Congress opposed President Reagan's July 1985 stockpile modernization recommendations, which would have reduced chromium inventories by 40 percent, cut cobalt and manganese holdings roughly in half, and eliminated platinum and palladium inventories completely. The current stockpile (see table A-6) contains a three-year supply of chromium and cobalt, a two-year supply of manganese, and less than a one-year supply of platinum-group metals.

The President's proposal was based on a 1984-85 NSC stockpile study that had been harshly criticized by the General Accounting Office for its methodology, its lack of sensitivity analysis, and its apparent intent to return excess revenues from disposals to the Treasury's General Fund in seeming violation of the 1979 Stockpile Act.[17] In President Reagan's proposal to Congress, about $3.2 billion in surplus items would be sold within five years. The Reagan administration argued that previous stockpile studies, upon which current stockpile goals were established, contained a number of basic errors and unrealistic assumptions regarding oil availability, essential civilian requirements, and domestic materials production. (See the White House Press Release, appendix D.)

Among the many opponents to the proposal within Congress was Representative Charles E. Bennett of Florida (see chapter 4 for a more complete statement of his views), who proposed under H.R. 1392 that authority and responsibility for the management of the stockpile be transferred to the Department of Defense (DOD) and that Congress itself determine by law future stockpile goals. During hearings in March 1987, DOD expressed opposition to this bill, as did the newly appointed Director of FEMA, while a former Director of GSA stockpile operations and a retired rear admiral expressed support for the bill.[18]

On 25 February 1988, in Executive Order 12626, President Reagan elected the Department of Defense to be responsible for the National Defense Stockpile. The Defense Logistics Agency will handle stockpile operations previously divided between FEMA and GSA. Development of the annual stockpile management plan will come from the Bureau of Mines and other agencies, in cooperation with DOD.

Although it is not surprising that stockpile policy has fluctuated widely from one administration to another, even President Reagan found it difficult to maintain a consistent policy. In 1981, the newly elected President ordered a "long overdue" purchase of cobalt to raise the stockpiled inventory.[19] In 1982 President Reagan declared, "The United States is a naval power by necessity, critically dependent on the transoceanic import of vital strategic minerals."[20] Early in 1985 Secretary of State George Shultz, in a major policy address, said that "South Africa is not a small island. It is a regional powerhouse endowed with vast mineral resources. . . ." Presumably he was speaking for the President. The Reagan administration was steadfast in attempting to maintain good relations with Pretoria while seeking political change in

South Africa through his policy of constructive engagement. It can be presumed that one consideration for remaining on good terms with the Botha regime was to assure a continued supply of key strategic minerals from that country or at least not to intentionally jeopardize that supply. Even Congress recognized this important dependency, including in its Comprehensive Anti-apartheid Act of 1986 provisions to exempt selected strategic minerals from any potential prohibitions on imported goods from South African parastatals (the "List of Ten").

An explanation of the administration's apparent flip-flop on the stockpile issue, in light of its yeoman efforts to strengthen the US defense program, may lie in the fact that the President wears many hats. He must attempt to balance the competing and often contradictory demands of multiple public interests. Thus, if balancing the budget is the overriding public-interest concern, one might expect to see cuts in other programs that could have an impact on strategic minerals availability such as federally funded R&D for materials research, and, perhaps, an even greater reluctance to provide subsidies to private industry for the development of domestic production capacity.

Elements in Congress, nonetheless, have remained steadfast against reducing stockpile levels, particularly with the situation in South Africa so volatile. Congress, although often frustrating the achievement of stockpile goals over the years by failing to allocate sufficient funds as recommended by the executive branch, has been reluctant to sell off inventories once material has been purchased. Part of this reluctance stems from seemingly less than astute stockpile management, which has on more than one occasion resulted in "buy-high, sell-low" transactions. For example, the last cobalt purchase for the stockpile—5.2 million pounds—was made at $15 per pound. Massive cobalt sales now would earn less than half this price based on 1988 producer and spot market prices.

Private Stockpiles as a Cushion

User stockpiles provide an important cushion against supply disruptions. Virtually every company that uses industrial materials maintains business inventories to ensure continuity of production and delivery in the event of supply problems. Although generally not large compared with the National Stockpile, private inventories are among the most efficient means of protecting against short-term supply cutoffs because they are located at the point of need and in the form, grade, and quality needed by the user to maintain production. Their size tends to vary according to several factors: (a) political risk assessments; (b) interest rates; (c) anticipated price trends for future inventory purchases; and (d) efficiency in managing those inventories. With computer-controlled inventory maintenance, low metals prices, and high interest rates, average inventories in the early 1980s tended to fall. However, with the onset of escalating political turmoil in South Africa beginning about 1983, private firms have undoubtedly increased their inventories of strategic minerals. Estimates by the Bureau of Mines, Commerce, and industry trade journals made for end-of-year 1987 indicate that private stocks range from a 4- to 8-month supply—in terms of consumption—for chromium, cobalt, manganese, and vanadium to a 15-month supply for platinum. However, when calculated in terms of imports from southern Africa, private stocks of chrome and cobalt provide a 7- to 8-month supply; for manganese, platinum, and vanadium, 12 to 18 months. For platinum, where markets are tightest and prices have been rising over the past 18 months, the stock situation is mixed. The US auto manufacturers, the largest single platinum users, currently have about a 12-month supply of platinum and palladium on hand. For rhodium, however, the industry is seeking to build stocks to a three-year level of consumption. The Japanese auto industry is believed to have about a two- to three-month supply of platinum and is estimated to have doubled its rhodium purchases from the USSR in 1986. In Western Europe, where platinum use for catalytic converters is still small, stocks are probably less than one month's supply but can be expected to rise with the stricter emissions standards that began to be phased in beginning in 1988.

Such inventory levels could get US industry through a short-to-medium term interruption, depending on which metals were involved. They would be inadequate, however, to sustain industrial production over a long period and would soon be depleted in the event of a surge in demand brought on by industrial mobilization for a multi-year conventional war. In the latter case, annual demand for defense needs alone could be expected to double or triple and private inventories would be grossly inadequate.

Private industry has strong feelings about the government's responsibility to maintain an adequate national stockpile and generally opposed Reagan's 1985 stockpile proposal. Acquiring and maintaining excess private inventories, which are nonproductive assets, is prohibitively expensive and runs contrary to the corporate view—which is to operate at a profit for the benefit of their owners or shareholders. This, in turn, contributes to employment, private spending and, through the multiplier effect, a more robust economy—benefiting the public at large. As an alternative, some have suggested that the government provide special tax incentives to encourage stock building by private industry. Proponents of such policies also argue that private industry can carry out a stockpile program much more efficiently than can the US Government.

The Need for Resource Sharing

Formalized resource sharing is another area where an improvement in strategic mineral policy could help reduce US vulnerability. The United States has only one such agreement— that being with Canada. As noted in chapter 1, Canada is rich in minerals, with nearly 70 percent of its mineral exports going to the United States. Moreover, it has not been as fully explored as has the United States. Based on mobilization agreements signed in 1949 and 1950, the two governments have agreed to "use the production and resources of both countries for the best combined results"; to "develop a coordinated program of requirements, production, and procurement"; and to "institute coordinated controls over the distribution of scarce raw materials and supplies."[21] It would appear prudent to pursue such a policy with Mexico and, perhaps, even Brazil.

The situation with our European allies is a bit more complex. Europe and Japan are even more dependent on imported strategic minerals than is the United States. During World War II, the United States shared many of its mobilization resources with the UK, France, and other allies. Presumably, in the event of a conventional war in Europe, our allies would expect us to share our strategic mineral stockpiles with them should materials shortages develop in their countries, and we would be likely to do so. Indeed, their needs could be considerable given the fact that their stockpiling efforts have been minimal to date.

The French stockpile acquisitions program, begun in 1975 for economic reasons in order to support French industry, is thought to contain a two-month average supply of 10 or 12 minerals. Chromium stocks are thought to be equal to about one year of consumption, platinum six months, and palladium three to four months. According to industry sources, France is changing the product mix of the stockpile, recently emphasizing strategic minerals. These sources estimate that France sold off about $150 million in lead, zinc, nickel, and copper during 1984-85. The UK stockpile program, initiated in early 1983 in response to the Falklands War, after many on-again, off-again debates within Parliament, is thought to be worth about $60 million and to contain cobalt, chromium, manganese, and vanadium in unknown but relatively small amounts. A stockpile sell-off program, announced in November 1984 for budgetary reasons, was suspended, apparently because of the increased risk of supply interruptions brought on by the worsening situation in South Africa. Sweden has had a materials stockpiling program since the mid-1930s, consisting of three stockpiles—wartime defense, peacetime emergencies, and stocks held by industry. As of October 1982, the total value of defense stocks of metals was about $93 million out of a total of $482 million. Known to be included are chromium, manganese, platinum, and vanadium. Quantities are secret but probably amount to six to eight months of consumption. Essentially the same metals are in the peacetime stock. Although the goal is about three months of consumption, quantities are well below targets. Industrial stocks are thought to be larger, encouraged by generous tax incentives.

In West Germany a plan to stockpile one year's consumption of chromium, cobalt, manganese, and vanadium was dropped in 1980 for budgetary reasons. Instead, West German policy is designed to encourage private enterprise to diversify foreign supplies, expand domestic minerals production, maintain technological leadership, expand recycling, and develop conservation and substitution technology in strategic metals. No other countries in Western Europe are known to be stockpiling strategic minerals. Judicious strategic minerals policy suggests that the United States ought to press the allies to do more in the area of stockpiling. The Reagan administration's proposal to drastically reduce US stockpile levels, however, could have sent the wrong signal to our NATO allies, making such negotiations difficult, if not impossible, to carry out successfully.

Although the absence of major stockpiles in Western Europe would preclude them as a source of raw materials for the United States, European metals refining and processing capacity is sizable, especially for the Big Four minerals. Much of the world's raw platinum and cobalt is processed in Europe and ferrochromium and ferromanganese capacity is sizable. According to 1984 data, West European production of all ferroalloys was some 2.5 times the level of US output with operating rates averaging only about 75 percent of capacity. The US industry, which was in sharp decline through 1986, depends increasingly on imports. During the 1978-85 period in the US ferroalloy industry, the number of plants declined from 29 to 17, and employment and capacity fell by roughly 50 percent. Imports now stand at roughly two-thirds of consumption. According to the US Bureau of Mines, January 1988 "posted the strongest month of raw steel production since 1984. . . . From January through July, the steel industry was operating at 88 percent of its capacity compared with only 77 percent a year earlier." One study concluded that during a mobilization, US imports would have to increase by 1.4 million tons to compensate for the decline in domestic capacity and a surge in demand.[22] This quantity is about equal to excess European capacity. Depending on the particular wartime scenario, European demand would also surge. Thus, an agreement to share any increased output of processed metals in exchange for raw materials in the stockpile could benefit both sides.

Diversifying Supplies from Abroad

Agreement is nearly unanimous that the United States ought to be diversifying its foreign supply sources in order to minimize the effects of a supply disruption from any one country. Although in theory such an approach is a good one, in practice such a policy is constrained by geological facts of nature. For the four most critical minerals—chromium, cobalt, manganese, and the platinum-group metals—known resources are heavily concentrated in southern Africa and the USSR. Nonetheless, a systematic approach to exploration in the LDCs might yield additional sources of supply. In general, these countries have not been thoroughly explored for strategic minerals due primarily to a lack of know-how and inadequate funds. Where the United States has provided aid and technology to LDCs, most efforts have been channeled into education and agriculture.

Past efforts along the lines of diversification have been piecemeal at best, but the institutions exist to carry out such a program. Organizations such as the Agency for International Development, the Trade and Development Program (TDP) of the International Development Cooperation Agency (IDCA), the International Program of the US Geological Survey, the US Export-Import bank, the Overseas Private Investment Corporation, the Export Trading Company Act, and Bilateral Investment Treaties have the authority to promote both government and private initiatives that could be used to support overseas mineral development.[23] In some cases, however, these agencies seem to work at cross-purposes with other aspects of the public good. For example, a $75 million loan by Eximbank to expand Mexican copper mining and smelting operations and a $35 million credit to expand the Mexican phosphate industry were viewed as harmful by similar US mining interests. Moreover, copper and phosphates are not considered to be strategic minerals. Despite these occasional incongruities, Eximbank programs now support about one hundred US mining and processing operations abroad.

In some cases, these agencies are constrained by other factors. For example, the Overseas Private Investment Corporation (OPIC) is required to consider the record of human rights before approving projects in developing countries. Similarly, amendments to the OPIC Act in 1978 made direct loan and feasibility study funds available specifically for the minerals sector, but the projects must be sponsored by US small businesses and expenditures cannot exceed $4 million—an amount hardly adequate to support capital-intensive mineral development projects.

The most successful initiatives in the past have been undertaken by the US Geological Survey. Major mineral discoveries have included manganese in Brazil, copper in Pakistan, potash in Thailand, and lithium in Brazil. Budget cutbacks have occasionally taken their toll, however. In 1969-70, the Geological Survey had 25 projects in 19 countries. By 1980, only two such projects existed. The Trade and Development program of the International Development Cooperation Agency has also been quite successful, supporting development feasibility studies of cobalt resources in Morocco and Peru, chromium in Turkey and the Philippines, and manganese in Mexico and Gabon—all strategic minerals. The Trade Development program, however, once constrained by budgetary limitations, received funding of only $700,000 in 1984; in 1988 the program received $25 million. Multilateral programs under UN or World Bank sponsorship have been the primary means of providing US assistance to the mining and smelting industries in the developing countries, but these programs, according to the GAO, have had only a limited impact on increasing supplies of strategic minerals considered to be important to the United States. Of 45 mineral-related projects during 1971-80, most involved copper, lead, zinc, and iron ore.[24]

The above discussion suggests that there exists a great deal of fragmentation in US overseas mineral development policy, a lack of well-defined objectives, and serious budget limitations. It would seem appropriate to combine all these efforts under one overseas program, taking advantage of economies of scale both in terms of budget allocations and US expertise. Perhaps the greatest benefit of such a reorganization of effort, however, would come from defining US objectives and pursuing them with a single-minded purpose. This would eliminate the impediment of attempting to achieve the public good across a wide range of foreign policy objectives. The payoff in new and diversified sources of strategic mineral supplies would likely be higher under such an arrangement.

Can Technology Provide the Answer?

Federal funding of research and development (R&D) provides another approach toward advancing the public good in strategic minerals, primarily through the development of substitute materials. Galvanized into action by the 1978-79 cobalt crisis and the realization that foreign supplies of other strategic minerals also could be easily jeopardized, serious research in the areas of technical conservation and substitution are now underway in the United States. Programs initiated by both government and private industry are investigating the feasibility of developing substitutes for strategic minerals and ways to use less material to perform the same function.

Private industry took the lead initially—first in cobalt, later in chromium. Substitutes were quickly developed in magnets and tool steels. Designing cobalt and chromium out of superalloys, however, has not been as simple a task, and has been considerably more costly. For example, INCONEL MA 6000—a cobalt-free superalloy was developed by a researcher at the International Nickel Company (INCO) in 1968, but its commercial introduction is probably five years away because of development costs in the $10 million to $12 million range. The private sector also has low-chrome and chrome-free alloys in the laboratory stage, designed to replace high-chromium stainless steel. The speed at which these new alloys become commercial will depend in large measure on whether there is a threat of a supply interruption—perceived or actual—and on prospects for chromium prices. At only about 40 cents per pound for ferrochromium, there is little price incentive to commercialize many of these laboratory developments. On the other hand, if chromium supplies should be disrupted—no matter how low chromium prices are—the incentive to replace them with other materials would build rapidly.

Because of the high costs and long lead times involved in developing substitutes, however, private industry is unable to carry the entire R&D load alone. Simply put, greater and more immediate payoffs are available through other uses of these funds. Therefore, the federal government has stepped in with its own long-range R&D substitution programs, working closely in many cases with private industry.

Development of another cobalt-free superalloy, underwritten by NASA, is expected eventually to cost $6-9 million—an amount no single company is likely to undertake for the purpose of contingency planning in the event of a hypothetical future cobalt shortage. Commercial development of the 9 percent chromium-molybdenum alloy mentioned earlier took nine years at a cost of $7 million—$5 million in government funding and $2 million from industry. In the area of structural ceramics, government-funded R&D has averaged $46 million a year since 1983.[25]

Some successes have been notable. Nearing commercialization is a 9 percent nickel-molybdenum alloy developed by Oak Ridge National Laboratory for use in the pressure vessels and connecting tubing of liquid-metal, fast-breeder reactors. This alloy uses 50 percent less chromium than standard stainless steel—which normally contains 16-26 percent chromium—and is now being considered for final certification by the American Society of Mechanical Engineers. NASA is developing a 12 percent chromium alloy steel to compete with the popular type 304 stainless, which is 18 percent chromium and accounts for 22 percent of total US chromium consumption. This alloy alone could reduce US chromium consumption by 15 percent. Similarly, the Bureau of Mines is investigating a chromium-free alloy to replace type 304 stainless.

Even the federal government is constrained, however, by the high costs of research and development of substitute materials. Funding for strategic materials must compete with other uses of the federal tax dollar. According to the Department of Commerce, only about $67 million or 4 percent of the nearly $1.7 billion federal materials R&D budget was directed at strategic minerals. In contrast $80 million went into the research and development of composites. For structural ceramics, federal R&D funding totaled $23 million in 1982, climbing to $40 million by 1984. Most of this total, however, went toward various forms of energy conservation or generation. Eventually there may be widespread uses for composites and ceramics in the military, aerospace, and automotive industries. Few of these new materials, however, are able to substitute for strategic minerals in their uses in current weapons systems.

In addition to the cost factor, institutional barriers often inhibit the speedy commercialization of substitutes in time of crisis. Some of these barriers could be surmounted by a technology-oriented government policy on strategic minerals. Lack of a centralized, readily accessible data base on material properties and structural analysis frustrates rapid design changes. Much private research is done in-house, and because results are generally proprietary, they are not shared with others working on the same problem. In a materials crisis, significant delays also could be caused by the need for qualification and certification testing. Without a ready commercial market or guaranteed government contracts, industry has little incentive to underwrite the expense needed to certify a new alloy or material. This process can cost from $20,000 to $200,000, depending on the scope of the testing, and take anywhere from six months to a year for military applications.

The plodding progress noted above reflects several other factors: (a) a lack of serious concern on the part of industry over the immediacy of any supply disruptions, (b) the potential for other cheaper sources of supply, and (c) the weakness of most markets for strategic minerals. Spot market cobalt prices, for example, have fallen from $11.00/lb. in 1984 to about $7.50/lb. in 1988. The trend in prices for chromium, manganese, and vanadium has also been downward. Only platinum prices have deviated from this trend, due primarily to speculative buying.

Growing Complexity in the Policy Process

The strategic minerals issue historically has not suffered from lack of attention, as shown in "The Hydranean Road to US Strategic Minerals Policy Formulation," which is by no means all-inclusive. Indeed, as the complexity of the issue has grown, more and more expertise has been brought to bear on the problem. Virtually every agency in the executive branch now has a department or office tasked with providing policy support on strategic minerals. The amount of activity devoted to the problem, however, has tended to run in cycles, depending on geopolitical events and the strategic awareness of the President in office. The fact that the strategic minerals foreign dependency issue still remains in a state of flux after years of study can be traced to several factors:

—fragmentation of authority;
—lack of coherence and continuity of policy;
—general disregard for the recommendations of independent study commissions;
—overall budget constraints and, within that, fragmentation of allocated funds; and
—loss of sight of the strategic minerals' public good because of competing equities among various government departments.

Some of the obstacles to a solution fall into the category of institutional constraints. Indeed the sheer size of the government may have become one of the most deleterious factors working against finding the best solution to the issues previously discussed. The ability of any one mind to have a major impact on the issues has been diluted by the need for compromise and the desire to hear all sides of the argument. On the other hand, the current system of inter-agency working groups assures that no single agency is able blatantly to pursue its own agenda or short-sighted policies to the exclusion of others.

Indeed, the complexity of the issues literally demands that all interested agencies weigh in with their own independent analysis. What may seem on the surface like a simple policy solution to one group is likely to have unknown or unintended implications in other areas—ranging from domestic jobs to international economic impacts and foreign policy repercussions. In order to capture these potential impacts, at least 10 agencies are now involved in strategic minerals analysis and policy formulation—Commerce, State, Treasury, Office of Management and Budget, the Defense Intelligence Agency, the Central Intelligence Agency, the Department of Defense, the Federal Emergency Management Agency,

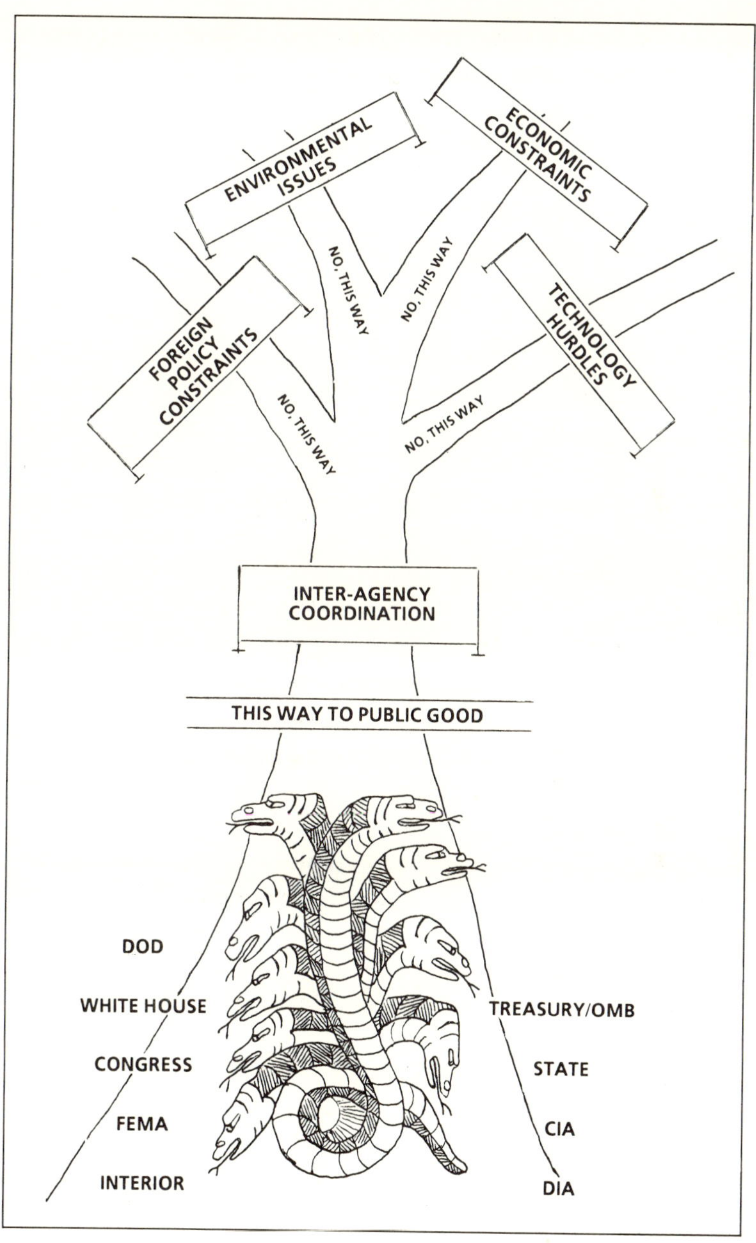

The Hydranean Road to US Strategic Minerals Policy Formulation.

the Interior Department, and the National Security Council—plus the White House and several committees of the Congress.

Continuing Stockpile Controversy

The year 1987 saw a number of new developments relating to the controversial NSC stockpile study and President Reagan's decision to drastically reduce currently held inventories. To recapitulate, at the Cabinet level the Secretaries of Commerce, Interior, Defense and State originally objected to the findings of the NSC study. Since then the GAO, US allies, domestic industry, and numerous trade associations have expressed their concerns in writing to Congress. As a result, Congress prohibited any changes in stockpile goals under the past three annual Defense Authorization Acts (1986-88).

This logjam has been further complicated by the fact that since the passage of the Comprehensive Anti-apartheid Act of 1986 sales of certain strategic minerals by the USSR and Eastern Europe to the United States have increased precipitously. For example, during late 1986 and early 1987, Soviet sales of antimony increased by more than 8,000 percent, chrome ore by more than 800 percent, platinum sponge by 157 percent, platinum bars by 724 percent, and rhodium by 390 percent over the baseline period of 1981-85. Similarly, for Eastern Europe, Czechoslovakia began selling platinum and palladium to the United States and Yugoslavia, ferrosiliconmanganese for the first time although neither country has any domestic resources of these ores. This increase in US dependency on the Communist countries has become an issue of some concern.

Politically, developments in 1987 included

—a request to the NSC by the Secretary of Interior, urging a reopening of the stockpile goals study;

—a policy statement by the National Strategic Materials and Minerals Program Advisory Committee urging that "a fresh start be made to provide seriously needed updating of stockpile plans;" and

—concurrence by DOD in the findings of the 1987 GAO report which harshly criticized the methodology, assumption and findings of the NSC study.

As a result of immense pressure and the congressional impasse against the stockpile disposals recommended by the White House, the NSC agreed in April 1987 to coordinate an interagency review of defense requirements for strategic minerals and industrial investment assumptions and to work with FEMA in identifying new materials that meet the criteria for inclusion in the stockpile. The latter point was particularly important to President Reagan's SDI program. A wide range of rare earth metals and advanced materials will be required to deploy the SDI system. Of these, only one—gallium—has been recommended to be purchased for the National Defense Stockpile.

Strategic Defense Initiative Materials Needs

Beryllium	Mercury	Selenium
Bismuth	Osmium	Strontium
Cesium	PGMs	Tellurium
Chromium (high-purity)	Rare Earths	Yttrium
Gallium	Rhenium	Zirconium
Hafnium	Rhodium	Metal Matrix
Indium	Ruthenium	Composites
Manganese (high-purity)	Scandium	and Fibers

In addition, SDI will need high purity forms of cobalt, columbium, nickel, tantalum, and titanium as alloying materials for rocket and structural uses. Curiously, the NSC study recommended large disposals of cobalt, tantalum, and titanium and the complete disposal of columbium, nickel, and three platinum-group metals—platinum, palladium, and indium.

A Need for Reorganization?

As the stockpile controversy vividly illustrates, each agency has, by nature of its charter, legal authorities and responsibilities for some aspect of strategic minerals policy formulation and analysis. Moreover, each has access to a certain body of information, not generally available to the others, that can have an important bearing on the subject. One could legitimately question whether the current system is an efficient way to do business, especially considering the chronology of tangled acts and agencies. It would seem that a strong case could be made for streamlining the current set-up by putting all the personnel now involved in strategic minerals analysis under one roof. Several advantages would accrue from a functional reorganization along such lines, the greatest of which would allow for speedier analysis and decisionmaking. A second major advantage would be the subjugation of multiple or hidden agendas to the pursual of a single agenda—the formulation and implementation of a coherent and integrated US strategic mineral policy designed to advance the goal of the public good. Such an approach could elevate strategic minerals issues to their rightful place. This point will be addressed in more detail in the concluding chapter.

4

Private Sector and Government Perspectives

PREVIOUS CHAPTERS EXAMINED AND ANALYZED THE SERIOUSness of US dependence on foreign supplies of strategic minerals to weigh the risks involved, some offsets to these risks, and the nature of US policy approaches to the problem. In order to provide a balanced analysis, some of the leading experts in the strategic minerals field—both in private industry and within the US Government—were asked for their perspectives on the subject of US strategic minerals vulnerability. This chapter presents this collection of essays, representing the viewpoints of the experts as they see the problem. Essays from the private sector are presented first, followed by those from the public sector. No attempt was made to influence the experts' positions. The goal was to elicit analysis, concerns, ideas, suggestions, solutions, and policy prescriptions. The author is indebted to the contributors for their spirit of cooperation and willingness to participate in this effort. Although admittedly only a sampling of outside opinions, it is felt that these views are nonetheless representative of how others may see the problem of US strategic minerals dependency.

Section I. The Private Sector

Platinum-Group Metals and Automotive Industry Demand

Kelly M. Brown, Michael J. Schwarz, and Dewain C. Belote, Ford Motor Company

Note. The following essay, prepared by the Ford Motor Company, discusses the potential for worldwide shortages of rhodium, one of the platinum-group metals. The auto industry accounts for about 60 percent of all rhodium used. While rhodium's use in catalytic converters cannot be considered strategic in and of itself, it is critical from a domestic economic standpoint. Moreover, any significant global supply shortfall could have an impact on its availability for strategic uses in the chemical, petroleum, ceramics, and electronics industries. In addition, the National Defense Stockpile contains no rhodium in its inventories. Finally, the only major alternative supplier outside South Africa is the USSR.

The automotive industry uses platinum-group metals (PGMs) for catalytic converters that reduce harmful pollutants from motor vehicles by means of chemical conversion of these pollutants in the vehicle exhaust. The installation of a catalyst in the vehicle exhaust system depends primarily on the stringency of the pollutant standards. For the United States, Canada, and Japan, the current allowable pollutant levels require catalysts on nearly all passenger cars and light trucks. Europe, Brazil, and Australia have pollution controls standards that require or will require catalysts.

There are basically two types of catalysts used for treating gasoline-powered vehicles:

—The conventional oxidation catalyst (COC), which contains platinum (Pt) and palladium (Pd), is effective for catalyzing the oxidation reactions of hydrocarbons (HC) and carbon monoxide (CO). The COC is most efficient with leaner than stoichiometrical fuel calibrations and secondary air supply.

—The three-way catalyst (TWC), which contains platinum (Pt) and rhodium (Rh), is effective for catalyzing simultaneously both the oxidation reactions of HC and CO and the reduction reactions of nitrogen oxides (NO). The TWC is most efficient with fuel calibrated within a narrow air-fuel ratio range around stoichiometry.[1]

With greater worldwide emphasis on control of vehicle exhaust pollutants, the anticipated use of catalytic converters using rhodium may exceed the supply of rhodium. Based on Ford estimates of PGM demand for regulated vehicles under three possible scenarios, a total worldwide demand for model year 1990 is projected.

US and Canada Demand

The projection of total PGM demand in 1990 for the United States and Canadian automotive industries is based on Ford's estimate of actual 1985 total industry usage and Ford estimates of PGM needs under various hypothetical future emissions standards. Only the nitrogen oxide standards were varied in the three regulatory scenarios that were analyzed. This was done for simplicity and because of the critical nature of rhodium—the rarest of the three PGMs. The scenarios are as follows:

Scenario A represents emissions standards that already have been adopted by federal and California regulatory bodies. Scenarios B and C represent more stringent standards that have appeared in various regulatory or legislative proposals. The estimated US and Canadian use of platinum, palladium, and rhodium under these scenarios is shown in table 1.

*Emissions of Oxides of Nitrogen**

PASSENGER CAR

	49 States (less Calif.)	*California*
1985 Base	1.0	0.7
Scenarios:		
A	1.0	0.4
B	0.7	0.4
C	0.4	0.4

* Grams per mile

LIGHT-DUTY TRUCK		
1985	2.3	1.0
Scenarios:	1.2/1.7**	.4/1.0
A	1.2/1.7	.4/1.0
B	0.5/1.7	.4/.5
C		

** Oxides of nitrogen (NO_x) standards are divided into two classes of light-duty trucks (LDT) based on loaded vehicles weight (LVW)—which is the vehicle curb weight plus 300 lbs. The first standard (for example, 1.2 gpm) is for LDTs with an LVW of 3,750 pounds or less, and the second standard (for example, 1.7 gpm) is for LDTs with an LVW of greater than 3,750 pounds.

Table 1. 1985/1990 Estimated Demand (US and Canada)*

	1985 Industry Use	*1990 Scenario "A"*	*1990 Scenario "B"*	*1990 Scenario "C"*
Platinum	687.0	684.0	749.5	834.2
Palladium	163.7	181.6	182.6	200.5
Rhodium	53.2	126.7	139.5	227.5

* Troy oz. × 1,000

Demand Outside the United States and Canada

Brazil, Australia, Europe and Japan have catalyst-forcing vehicle pollution emission standards. Ford's estimates of PGM usage required to meet these standards are shown in table 2. The confidence in these estimates is not as high as for the United States and Canada because the standards in Australia, Brazil, and West Germany could be met without catalysts or by using a unique, precious-metal-loading technique if catalysts are used. The best estimate for 1990 is shown in table 2.

Table 2. 1990 Estimated Auto Demand (Outside US and Canada)*

	Platinum	*Palladium*	*Rhodium*
Brazil	26.7	8.3	1.0
Australia	25.6	6.5	5.8
Europe	78.8	15.5	12.7
Japan	158.0	35.2	29.4
Total	289.1	65.5	48.9

* Troy oz. × 1,000

Total Free World Demand

Total Free World Demand (table 3) is the sum of US and Canada demand (table 1) and demand outside the US and Canada (table 2). The platinum:rhodium demand ratio in converters is departing further from the current South African mining ratio of about 19:1. The 1985 platinum:rhodium demand ratio for the industry was 10:1, and the projected 1990 demand under scenario C assumes a continued decline to a ratio of 4:1. With the development of the rhodium-rich UG2 reef, which has an estimated mining ratio of 5:1, the overall mining ratio for existing mines plus UG2 mines will be reduced.[2] Thus, the mining of platinum and rhodium should in time become somewhat more compatible with demand.

Table 4 lists total PGM demand for 1984 and 1985 along with the auto demand for the total Western world as reported by Johnson-Matthey in their "Platinum 1985" and "Platinum 1986" publications.

Comparing the Ford estimate of total auto industry usage for 1985 in table 3 to that reported by Johnson-Matthey in table 4, the estimated usage for each metal is less than that reported by the industry. Some of the difference could be attributable to inaccuracies in estimating or reporting. In any event, it is likely that some rhodium is being stockpiled by the industry.

Table 3. 1985/1990 Estimated Demand (US, Canada and Rest of World)*

	1985	*1990 Scenario "A"*	*1990 Scenario "B"*	*1990 Scenario "C"*
Platinum	845.0	973.1	1038.6	1123.3
Palladium	199.0	247.1	248.1	266.0
Rhodium	82.6	175.6	188.4	276.4

* Troy 0.2 × 1,000; for US and Canada, figures are autos and trucks; for rest of world, figures are autos only.

Table 4. Western World Demand*

	Total Demand		*Auto Demand*	
	1984	*1985*	*1984*	*1985*
Platinum	2,590	2,810	725	875
Palladium	3,100	2,740	340	290
Rhodium	224	250	110	135

* Troy oz. × 1,000

Because the Ford estimate of world rhodium demand for the automotive industry in 1985 is substantially lower than that reported by Johnson-Matthey (826 versus 135 thousand troy ounces), it may be prudent to estimate rhodium demand as ranges under the three 1990 scenarios. The range for each scenario is bounded on the low side by the Ford projections of automotive demand from table 3. The upper limit is calculated by applying the ratio of the estimated demand for each scenario over the 1985 base (from table 3) to the Johnson-Matthey reported 1985 auto-sector demand of 135,000 troy ounces. In both cases non-automotive demand for rhodium is held constant at 115,000 troy ounces. Table 5 thus shows the impact of projected increases in automotive demand on total demand. On the low side, total demand is projected to increase by 16 to 57 percent by 1990 for

Table 5. Projected Free World Rhodium Demand*

	Lower Limit		*Upper Limit*	
Scenario	*Auto Demand*	*Estimated Total Demand***	*Auto Demand*	*Estimated Total Demand***
1985 Base	82.6	250	135	250
1990 A	175.6	291 (16%)***	287	402 (61%)
1990 B	188.4	303 (21%)	308	423 (69%)
1990 C	276.4	391 (57%)	452	567 (127%)

* Troy oz. × 1,000
** Base total demand = 250,000 troy ounces, consisting of 135,000 automotive and 115,000 non-automotive.
*** Percent increase over 1985 base demand.

the three scenarios; on the high side, projected increases range from 61 percent to 127 percent.

Supply of Platinum and Rhodium

The two major South African PGM producers (Rustenberg and Impala) appear to be working near capacity. Estimates of the supply of mined platinum and rhodium for 1985 were slightly less than demand. This deficit was made up from supplies existing in the marketplace.

Although there have been no signs of expansion of mining capacity by the major producers, it is felt that they are waiting to determine actual industrial demand, absent current speculative trends in the platinum market, before committing to the large capital cost of expansion.[3] It is generally estimated that once the commitment is made, approximately five years will be required to add any significant mining capacity. On the other hand, a disproportionate increase in rhodium demand might be met partially by increased mining of the UG2 reef (platinum:rhodium 5.25:1) compared to the Merensky reef (platinum:rhodium 19.7:1). On balance, it appears that mining capacity would be inadequate in 1990 under Scenario C of the lower limit projections or any of the three scenarios of the upper limit projections.

Future Technology

The automobile industry is striving to develop low pollution emitting engines so that the use of catalysts can be reduced or the amount of precious metal used can be reduced. Other developments such as methanol fuel vehicles would change the type of catalyst required and alter the amount of precious metals needed. These technological changes most likely will not occur prior to 1990. Major uncertainties in availability, pricing, and the source (South Africa) of rhodium will provide the stimulus to control engine-emitted pollutants with modified catalysts, striving to reduce the amount of platinum group metals, including rhodium.

Conclusion

It appears that without further changes in automobile emissions standards (Scenario A), the demand for rhodium should increase by 16 to 61 percent. If US standards are tightened to stringent levels contained in certain proposed legislation (Scenario C), the demand for rhodium could increase by 57 to 127 percent, raising serious doubt as to the ability of mining capacity to expand sufficiently to meet demand.

With a moderate increase in PGM production and the readjustment of other demands, platinum and palladium supplies should be adequate to meet 1990 demand.[4] In fact, if rhodium demands of the nature discussed above are to be met, a substantial oversupply of platinum and palladium would occur, based on expected mining ratios.

Notes

1. Stoichiometric refers to the air-fuel ratio for complete fuel combustion (in the case of gasoline, the average ratio is 14:1, i.e., 14 parts of air to 1 part of gasoline).

2. The UG2 reef is a deep ore body in South Africa, rich in chromium and platinum group metals, that is still in the early stages of development because of its complex mining problems and higher associated costs of production. Significantly greater PGM output can eventually be expected from this ore body, in part because it is rich in rhodium.

3. A third, relatively minor South African producer is Western Platinum. Another company, Gold Fields of South Africa, is currently evaluating its reserves and could begin PGM production in the early 1990s, depending on market conditions.

4. Other industry demands include chemical, electrical, glass, jewelry, and petroleum.

Kelly M. Brown, a Ford Motor executive engineer, is responsible for fuel economy, emissions and noise-related regulatory matters involving government interaction and company planning and represents Ford on various committees of the Motor Vehicle Manufacturers Association and the Engine Manufacturers Association.

Michael J. Schwarz is the Manager of the Emission Control Analysis and Planning Department of Ford Motor Company. His work in the areas of vehicle emissions and fuel economy has involved certification compliance and regulatory planning.

DeWain C. Belote is an Emissions Planning Associate of the Environmental and Safety Engineering Staff of Ford Motor Company. Mr. Belote is responsible for preparing Company statements and responses to government agencies and contractors on emissions matters and for developing forecasts of future emissions regulations.

Toward Strategic Metals Independence in Jet Engine Manufacturing

Eugene Montany

Note. The following article by Eugene Montany, Vice-President for Technology and Strategic Planning at Pratt & Whitney, underscores the importance of strategic metals in the production of jet engines—both military and civilian. In the article, Mr. Montany stresses the need for better stockpile planning, a stronger R&D effort in developing substitutes, and the need for the private and government sectors to work together more closely. The article also details the internal work that Pratt & Whitney is doing in the areas of improved manufacturing technologies, conservation recycling, and materials substitution.

Pratt & Whitney, a division of United Technologies Corporation, is a leading supplier of jet engines for commercial airlines and for the military. None of our engines can be manufactured today without raw materials that are classified as strategic because of their importance to America's national security.

Many complex factors influence the strength of the defense industrial base and the nation's ability to gear up military production to generate additional surge capacity. One critical factor is the availability of strategic raw materials. About 13,000 pounds of metal go into the manufacture of a jet engine such as Pratt & Whitney's F100, which powers the F-15 and F-16 fighter aircraft. Among the metals required are chromium, cobalt, columbium, nickel, tantalum, and titanium. Although the cost of the base raw materials is only a small percentage of the total cost of building an engine, modern high-performance engines cannot be built without them. The United States has virtually no primary production of chromium, cobalt, columbium, tantalum, or manganese and only minimal production of titanium and platinum-group metals. All are integral to the manufacture of aerospace equipment. All must be imported to meet US demand.

The degree of foreign dependency is cause for concern in and of itself, but it becomes especially alarming when one considers the countries on which we are dependent for the supply. The Republic of South Africa is our primary supplier for several of the materials critical to the industrial base and national security, including chromium. South Africa, moreover, is the major transshipment point for other minerals mined in southern Africa, such as the cobalt from Zaire used in aircraft engines. According to a July 1985 Commerce Department assessment, continued and unrestricted access to South African mineral supplies is "vital for the continuing [US] defense buildup and for industrial preparedness in the event of a national security emergency."

If supplies from South Africa were curtailed, the most promising alternative supplier would be the Soviet Union—a sobering reality. Together, South Africa and the Soviet Union account for more than 90 percent of world platinum-group metals production, 56 percent of world chromium production, and 58 percent of total manganese production. The concentration of these important resources in the hands of the Soviet Union and South Africa, and our dependence on South Africa to meet current requirements, emphasize US vulnerability to a supply disruption. Because a single nation supplies the United States with such a large amount of its needs, it would be inordinately difficult to find a single new supplier to pick up the slack in the event of a supply disruption.

Although critical and strategic materials are currently exempt from the South African trade embargo imposed by the United States, there is growing cause for concern regarding the future supply of these materials. The South African government has threatened to curtail the supply of minerals in retaliation against US sanctions. Even if no preemptive decision to cut off supply is made, there remains the possibility that internal chaos, such as widespread mining strikes and crumbling of the South African infrastructure—including the vital rail system—could force *de facto* supply disruptions.

If supply were disrupted, the Soviet Union probably would delight in the opportunity to fill the void, as it did for cobalt in 1978 when Angola invaded Zaire; prices, of course, would soar. The USSR naturally could cut us off whenever it wanted to, for obvious reasons; there is no comfort in being dependent on the Soviets for materials vital to our national defense. The United States' critical materials safety net is supposed to be the National Defense Stockpile, but the stockpile is alarmingly deficient in terms of both quantity and quality of reserves for chromium, cobalt and the platinum-group metals.

Over the years, Pratt & Whitney had advocated a three-pronged strategy to address US vulnerability resulting from our minerals dependency. First, domestic production must be encouraged whenever economically and environmentally feasible. Second, R&D efforts must continue and lead to more substitution, conservation, recycling, and reduced requirements for critical and strategic materials. And third, the National Defense Stockpile must maintain an inventory of materials in adequate quantities, of sufficiently high quality, and in the appropriate form to supply US needs in the event foreign supplies are curtailed. For its part, the US Government recognized a long time ago that the United States was not self-sufficient in all the materials required for our national defense. The National Defense Stockpile was created to put aside materials which are not produced in sufficient quantities within our borders to satisfy our defense production and essential civilian needs. The stockpile currently maintains inventories of the materials we import from South Africa. Unfortunately, US vulnerability to supply disruption is compounded and exacerbated by the inadequacies of the stockpile. The stockpile suffers from serious problems in three areas: quantity, quality, and management.

With respect to quantity, the inventory of the stockpile contains only $8.3 billion worth of materials that are being held against a total stockpile goal value of $16.6 billion. Almost $2 billion of the inventory, however, represents excess materials, i.e., quantities in excess of their stockpile goals. To meet the goals would require acquisition of additional materials valued at approximately $10 billion. The cobalt inventory, for example, falls 32 million pounds short of the cobalt goal; that represents almost a 40 percent shortfall. In spite of these shortfalls, the administration proposed to Congress that the stockpile inventory be shrunk significantly through an ambitious program of disposals.

The inventory shortfalls are compounded by quality problems. Many of the stockpiled materials have deteriorated or have been rendered obsolete by technology or are not in the appropriate form to be of any use to the defense industrial base in a period of national emergency. For example, recent American Society of Metals reports on stockpiled cobalt and chromium reveal that only 10 percent of the cobalt and none of the chromium is usable for the highest grade vacuum-processed superalloy use while the balance, 40.8 million pounds purchased from 1947 to 1961, is not. The 7.5 million pounds of stockpiled chromium metal purchased in the 1960s, likewise, is not suitable for today's vacuum-processed superalloys.

Furthermore, only two US companies produce pure chromium metal to our specifications, and they have the capacity for only 50 percent of the total US demand. The balance must be imported. The 40.8 million pounds of unacceptable cobalt could be used in less exacting applications, but it would require upgrading or reprocessing before it could be used in more sophisticated end products such as the hot section of our F100 fighter engines. There is very limited capacity within the United States to upgrade the old cobalt stockpile, and, in an emergency, it would be time-consuming and risky to ship the material overseas for upgrading and then back to the United States for use.

The United States needs the National Defense Stockpile to ensure critical and strategic materials availability. Without a strong, functioning materials stockpile, the nation's security would be jeopardized. In light of the current stockpile deficiencies, Pratt & Whitney believes reform of national defense stockpile policy is mandatory.

Stockpile goals and inventories must be balanced. In determining quantities of materials to stockpile, stockpile managers must take into account this country's net import reliance for each given material. Other important factors such as reliability of supply or, conversely, vulnerability to supply disruption and availability of substitutes, for example, also must be considered. Once goals have been established, those goals should be met, and in a manner that minimizes market disruption for either producers or users of the materials.

Stockpile materials should not be stored in a form that requires further processing before they can be used or that does not meet industry standards. An upgrading program for cobalt currently in the stockpile is urgently needed. Increased consultations with industry are necessary to ensure that current and future stockpile purchases meet industry specifications, are in the necessary form, and are of sufficient quality for today's—and tomorrow's—sophisticated applications. Pratt & Whitney is an eager and active participant in this essential dialogue.

The stockpile acquisition and disposal program needs to be continued and accelerated if we are to bring stockpile goals and inventories into balance. The Stockpile Transaction Fund should, therefore, retain its independent status and continue as a revolving fund, with proceeds from inventory sales used solely for acquisitions or upgrading, *not* for balancing the federal budget. The stockpile should be isolated as much as possible from the political, economic, and budgetary pressures that have plagued stockpile management in the past and contributed to the stockpile's current deficiencies.

For our part, Pratt & Whitney has pursued and continues to pursue a number of internal programs to reduce our own dependence on critical and strategic materials. Our strategy consists essentially of the following components: improved manufacturing technologies, conservation, recycling, and material substitution. We also have sought to reduce requirements through repair and restoration of worn or damaged parts, thereby extending the service life of our equipment, and through the initiation of new designs with reduced part count and extended component durability.

Pratt & Whitney, for example, is developing metallic and ceramic protective coatings that not only are lower in strategic material content but also allow use of new or existing alloys that are "lean" in strategic elements. We are extending component life through improved nondestructive evaluation (NDE) and retirement for cause (RFC) criteria. A study conducted on 21 F100 rotating parts showed that 3,050 short tons of strategic materials could be saved over the life of the system through the implementation of RFC methodology.

Manufacturing technology (ManTech) programs under Air Force sponsorship also have helped to reduce the input requirements in component manufacture. As a result of a ManTech program to scale up conventional casting techniques to large cases, our manufacture of large precision titanium castings reduces input material more than 50 percent relative to cases fabricated from forgings. And large precision nickel-base superalloy castings use 50 percent less input material than cases fabricated from forgings. The Gatorizing® forging process developed by Pratt & Whitney results in a 50 percent reduction in input material in a typical F100 disk that, in part, had been produced by conventional hammer- or press-forging, and refinements to the Gatorizing® process allow a 40 percent reduction in input material for nine F100 components.[1]

In the area of materials substitution, a good deal of research and development has focused on the potential for replacing superalloys with ceramics. Pratt & Whitney is using six small ceramic parts in one of its engines. This application represents the first time a ceramic part has been used in commercial production.

We are very proud of the results we have achieved at Pratt & Whitney. The introduction of new materials and technologies, however, can take considerable time—about 5 to 10 years. This was clearly articulated in a January 1985 Office of Technology Assessment report entitled "Strategic Materials: Technologies to Reduce US Import Vulnerability." Despite our ongoing efforts, for instance, Pratt & Whitney's cobalt and chromium consumption is not projected to begin decreasing until 1989. And it will be a long time before ceramic parts have widespread commercial use in turbine engines.

It should be clear that this nation needs the National Defense Stockpile to ensure critical and strategic minerals availability. At the same time, we need to strengthen our industrial base by encouraging the modernization and expansion of domestic production, processing, and conservation of critical and strategic materials. Finally, industry and government must work together to focus on strategies which will ensure that adequate supplies of strategic and critical materials are available to meet national defense and essential civilian requirements.

Notes

1. The Gatorizing® forging process, in general, is a hot-die forging process. In this process, the alloy to be forged is placed, by particular thermomechanical processing, in a temporary condition of "superplasticity," i.e., low-strength and high ductility, and forged to the desired configuration under hot isothermal conditions, meaning the dies and forging stock are heated to the same required forging temperature and maintained at that temperature during forging. Subsequent to the forging operation, the alloy is returned to a normal condition of high strength and hardness by heat treatment.

Eugene R. Montany, Vice President-Technology and Strategic Planning, Pratt & Whitney, is responsible for analyses and recommendations concerning strategic short- and long-range business and technology plans.

Dependence of Alloy Producers on International Strategic Minerals

Art M. Edwards

Note. The following article by A. M. Edwards, Communications and Advertising Manager for Haynes International, Inc., provides in technical detail the importance of chromium and cobalt in the manufacturing of high performance superalloys. The author stresses, in particular, that chromium is vital to the production of nearly all alloys produced by Haynes, that there are no adequate substitutes available for chromium, and that Haynes is virtually 100 percent dependent on South Africa for this material. Regarding cobalt, Mr. Edwards details the long and arduous substitution program that has enabled Haynes to reduce but not eliminate the need for cobalt in superalloys. The criticality of both of these materials is highlighted by their use in a number of strategic weapons systems including the F-15 "Eagle" fighter aircraft, jumbo jets, helicopters used in Vietnam, and the Apollo XIV Lunar Mission.

Cobalt and chromium are both of vital importance in the production of high-performance alloys at Haynes International. Cobalt is indispensable as a base metal for a number of alloys. Cobalt-base, solid-solution-strengthened alloys have intrinsically better high-temperature strength and hot corrosion resistance than similar nickel-base alloys. In nickel-base precipitation-strengthened alloys, cobalt also reduces the solubility of the "gamma prime" phase. This enhances its high-temperature strengthening capability. Practically all of the alloys produced at Haynes rely on chromium to provide high-temperature strength plus oxidation, hot corrosion, and aqueous corrosion resistance. Vanadium, manganese, titanium, and platinum are considerably less important.

During the 1960s, Haynes investigated the potential of vanadium metal as a commercial material. At that time, Union Carbide Corporation was a parent company and vanadium produced in the corporation's mines in Colorado and California as a uranium by-product was a glut on the market. A pilot plant operation, in which vanadium pentoxide was reduced with calcium metal, was operated for a number of months at Kokomo. After enough metal was produced to develop a meaningful properties profile, it was determined that the metal had about the same characteristics as bronze, an alloy that has served mankind well for over four millennia. Thus, the project was dropped.

Cobalt and Chromium

The connection between cobalt and Kokomo, Indiana, is the result of a long series of coincidences dating back to the turn of the century. In 1904, rich silver ores were discovered in northern Ontario. Unfortunately, the ores also contained arsenic, nickel, and cobalt as impurities. The fields remained untapped until 1907 when M. J. O'Brien, a prominent Canadian railroad entrepreneur, formed Deloro Mining and Reduction Company. His company utilized a new refining process developed at Queens University in Kingston, Ontario, to produce silver and white arsenic. In 1910, cobalt production was also started. The only drawback was that there were no significant markets for the metal except some small use of the oxide as a pigment in ceramics.

In the same year Elwood Haynes, a pioneer American automobile manufacturer and ardent metallurgist, delivered a paper at an American Chemical Society conference in San Francisco, entitled, "Alloys of Nickel and Cobalt with Chromium." Haynes' paper was reviewed in an Australian newspaper. An English newspaper reprinted the account and this second-hand report was, in turn, read by a Welsh associate of Thomas Southworth, a Deloro official. The friend promptly got in touch with Southworth in Canada who immediately contacted Haynes back in the United States and the circle was completed. Haynes now had a reliable source of cobalt for his alloys and Southworth had a market.

In 1912, Haynes founded the Haynes Stellite Company in Kokomo, Indiana, and started producing cobalt-base alloys at an ever-increasing tempo. Their major use was for highly efficient metal-cutting tools, although there was some use also for cutlery and surgical instruments. A similar enterprise, affiliated with Deloro, was started in Canada to serve the British Commonwealth. The two companies were merged in 1980 but separated in 1987. Since 1912, they have probably produced more cobalt alloys than the rest of the world combined. Canada remained the main source of the metal until 1925 when more economical deposits in the

Belgian Congo (now Zaire) came on stream. Since then, Haynes has been mainly dependent upon African sources. Today, about 75 percent of the cobalt consumed at Haynes is of African origin. Canada supplies 20 percent as a by-product of nickel production, and the remainder comes from Japan, also as a by-product.

As with most other metal companies, Haynes is virtually 100 percent dependent upon South Africa for chromium raw materials. Turkey and the USSR are potential, albeit uncertain, alternate sources. The situation is alleviated somewhat by the fact that raw materials supply less than one-half of the feedstock for high-performance alloys at Kokomo. Both process and purchased scrap make up over half of most furnace charges. The other side of the coin is that there is no acceptable substitute for chromium in imparting oxidation and corrosion resistance to high-performance alloys. Functional alloys can be produced with little or no cobalt but not without chromium.

Interestingly enough, it was an effort to utilize high-temperature alloys without chromium that launched Haynes into the superalloy business. In the late 1930s, the US Air Force (then known as the US Army Air Corps) funded a program to produce superchargers for aircraft engines from Hastelloy alloys A and B. Both were corrosion-resistant, nickel-base alloys that had been developed to handle hot hydrochloric acid, a chemical that can eat through stainless steel in short order. Neither Hastelloy alloy contains chromium, but both have acceptable oxidation resistance up to 1,400 °F. Alloy A, which contains about 20 percent iron and 15 percent molybdenum, was used for supercharger disks while the 28 percent molybdenum, 5 percent iron, Alloy B was used for blading. This combination worked, but ultimately the inherent superiority of cobalt-base alloys, containing chromium, won out and advanced blading was produced from cobalt-base alloy 21 (Co-28Cr-5.5 Mo-24 Ni).

Uses of nickel-molybdenum alloys at high temperatures still survive as evidence of some slight independence from chromium. They are used principally because of their low coefficients of thermal expansion. Hastelloy<R alloy B seals are used in a number of jet engines including the General Electric J79—one of the

largest volume military gas turbine engines ever produced. It powers the McDonnell Douglas F-4 Phantom jet. Alloy B is also used in another large volume GE military engine, the TF39, as well as the air-launched cruise missile engines (ALCM), produced by Williams International. Seven components in the space shuttle main engine are made of alloy B. Alloy B rocket nozzles were also used on the Viking Mars landers, but the alloy's unequaled high-temperatures applicability was not essential since there is very little oxygen in the Martian atmosphere. Despite these few minor exceptions, high-performance alloys for both high-temperature and corrosion-resistant applications would be in a very sorry state without chromium.

Types of Products Made

Nickel- and cobalt-base, heat- and corrosion-resistant alloys are produced at Haynes in the forms of sheet, plate, bar, billet, wire, welding products, and both welded and seamless pipe and tubing. The company no longer produces castings or hard-facing products. Some of these latter forms will also be described here since they, too, depend heavily upon cobalt and chromium for their major characteristics. The manifold uses of chromium pervade not only the cobalt alloys but also the enormous universe of nickel and iron-base (including stainless steel) alloys.

Aside from the previously mentioned Hastelloy alloy B, only high-nickel alloy 400 (67Ni-31Cu), titanium, and the relatively expensive refractory metals, such as tantalum and columbium and their alloys, have a chance of replacing chromium-bearing alloys for aqueous corrosion service. At one time, Haynes produced a nickel-silicon-copper alloy, called Hastelloy alloy D, which had good resistance to boiling sulfuric acid, but its usefulness was limited since it could be produced only in cast form. It was ultimately replaced by tantalum in its major application as sulfuric acid concentrator tubes.

Aerospace

By far, the largest use of cobalt is in nickel-base alloys for gas turbine engines. As mentioned previously, cobalt enhances the

Table 1. Gamma Prime Alloys[1]
(Contents in percent)

	Ni	*Cr*	*Co*	*Ti*	*Al*	*Mo*	*C*	*Fe*	*Others*
Cast alloys									
B-1900	Bal	8.0	10.0	1.0	6.0	6.0	0.11	—	Ta = 4.3 B = 0.015
713C	Bal	13.5	—	0.9	6.0	4.5	0.14	—	Zr = 0.10
Wrought alloys[2]									
Waspaloy alloy	Bal	19.5	13.5	3.0	1.25	4.25	0.07	—	B = 0.005
Rene 41 alloy	Bal	19.0	11.0	3.1	1.5	10.0	0.09	5.0	B = 0.006
Alloy No. 263	Bal	20.0	20.0	2.2	0.5	5.9	0.06	—	—

[1]The symbols used in tables 1 and 2 are standard "abbreviations" from the periodic table of the elements. They represent the following: Ni (nickel); Cr (chromium); Co (cobalt); Ti (titanium); Al (aluminum); Mo (molybdenum); C (carbon); Fe (iron); Ta (tantalum); B (boron); Zr (zirconium); Cb (columbium); Mn (Manganese); W (tungsten); and La (lanthanum).

[2]Hastelloy, Haynes, and Multimet are registered trademarks of Haynes International. Waspaloy is a trademark of United Technologies. Rene 41 alloy is a registered trademark of General Electric.

stability of a strengthening feature in alloys called "gamma prime." Some typical wrought "gamma prime" alloys, such as Waspaloy alloy and Rene 41 alloy are listed in table 1. These alloys are indispensable for engine vanes and blading, disks, seals, casings, shafts, and virtually all hot-section components. Replacing or modifying these alloys in the event of a disruption or severe shortage of cobalt would be possible but with performance penalties. This could not be done without chromium.

In addition to the use of cobalt in gamma prime alloys, cobalt-base alloys also play a large role in jet aircraft engines. Some compositions for typical cobalt-base alloys are given in table 2. Haynes alloy 188, which contains about 40 percent cobalt, is the major wrought alloy used.

Alloy 188 combustors gave the Pratt & Whitney F100 engine a 300 °F temperature advantage over existing hot section alloys. Since power and speed are directly related to the turbine inlet temperature, this was a major breakthrough. The F100 is the power plant for the McDonnell Douglas F-15 "Eagle" fighter aircraft

Table 2. Cobalt Base Alloys

Alloy	*Co*	*Ni*	*Si*	*Fe*	*Mn*	*Cr*	*Mo*	*W*	*C*	*Others*
No. 6b	Bal	3.0*	2.0*	3.0*	2.0*	30.0	1.5*	4.5	1.2	—
No. 6K	Bal	3.0*	2.0*	3.0*	2.0*	30.0	1.5*	4.5	1.6	—
No. 3PM***	Bal	3.0*	1.0*	3.0*	2.0*	31.0	—	12.5	2.3	Al = 1.0,* B = 1.0
No. 25	Bal	10.0	1.0*	3.0*	1.5	20.0	—	15.0	0.1	—
No. 31	Bal	10.5	1.0*	2.0*	1.0*	25.5	—	7.5	0.5	—
MP35N****	35.0	35.0	—	—	—	20.0	10.0	—	—	—
WI-52*	Bal	—	—	1.8	—	21.0	—	11.0	0.45	Cb + Ta = 2.0
MAR-M509**	Bal	10.0	0.1*	1.0	0.1*	21.5	—	7.0	0.6	Ti = 0.2, B = 0.1* Zr = 2.25, Ta = 4.5
No. 188	39.0	22.0	0.35	3.0*	1.25	22.0	—	14.0	0.10	La = 0.04
No. 150	Bal	3.0	0.35	20.0	0.65	28.0	1.5	—	0.08	—

* Maximum
** Cast alloy
*** Powder metal alloy
**** Registered trademark of SPS Technologies, Inc.

which has broken nearly every existing rate-of-climb record. The PWA F100 is also used on the newer General Dynamics F-16 "Falcon" single-engine, air-superiority fighter aircraft.

Haynes alloy 188 is also used for combustors for air- and sea-launched cruise missiles and in a number of auxiliary power units (APUs). Alloy 188 serves as the combustor material in the giant General Electric CF6-80 engine that powers many of the world's jumbo jets.

A total of 47 separate components of the space shuttle's main engine are made of Haynes alloy 188. Here the alloy was selected not only for high-temperature strength but also for its superb resistance to hydrogen embrittlement at cryogenic temperatures.

Other Uses

Although well over three-quarters of the cobalt alloys used in the United States go into gas turbine engines, there are many other important uses of cobalt alloys outside of the gas turbine field. Cobalt alloys have very high reflectivity, especially in the infrared range. The alloys were used with great success for naval searchlight reflectors during both world wars. Although their reflectivity is slightly inferior to silver, the resistance to tarnishing is far superior, especially in sea air. Cobalt alloy reflectors soon outshine their silver counterparts in service afloat. Cobalt alloys are still used in small quantities for reflectors in sextants, radiometers and other optical instruments.

Cobalt alloys also have an unsurpassed resistance to a phenomenon known as cavitation erosion. Cavitation erosion occurs on the surface of very high-speed components operating in fluids such as water, mercury and liquid sodium. The theory is that cavitation erosion is caused by tiny bubbles collapsing on the metal surface with great rapidity. One prime example of the problem is at the condensing end of steam power turbines in electric power plants. The solution for many years has been to inlay, or braze, metal sheet of a 58Co-30Cr-4W-1C alloy, known as Haynes alloy 6B on the leading edge of turbine blading. Practically every condensing-type turbine in the Western world is so outfitted. High-speed pumps and air- and hydrofoil surfaces can be protected in a similar manner.

During the Vietnam conflict, helicopter main rotors and tail rotors were also fitted out with alloy 6B erosion shields. The cobalt alloy protected the metal from the scouring of abrasive propwash when the copters landed and took off from unpaved strips in remote areas. Unprotected rotors usually failed in less than 500 hours in this service, but the cobalt alloy gave many times this life.

An interesting sidelight that illustrates the strategic importance of alloys occurred in 1966. With a shortage of these erosion shields imminent in December of that year because of a strike at the Kokomo plant, President Lyndon Johnson invoked the Taft-Hartley Act to order management and workers back on the job to assure a continuing supply of these and other alloy helicopter components. Historically, this is the only instance where action of this nature was taken by a chief executive.

Cobalt alloys also have a small but important spot in the medical field. Cobalt alloys do not cause clotting in blood as readily as other stainless alloys. Alloy wire and bar are used in heart valves and many prosthetic devices for this reason and for durability, strength, and compatibility with human tissue.

Unusually severe tests were carried out to select the biological container for plutonium-238 fuel used in radioisotope-fuel capsules for space instruments. Because of its very long half-life, plutonium-238 can supply a virtually unending supply of heat energy. However, its toxicity is an enormous threat. In selective testing for containment reliability, Haynes alloy 25 capsules were heated to 1,050 °F, accelerated on a rocket sled, and slammed against granite slabs. Even after exploding the grapefruit-sized capsules in TNT, they would not rupture or leak. Capsules such as this were part of a number of scientific and navigational packages in space. One is now on the moon as part of the Advanced Lunar Scientific Experiments Package (ALSEP), left behind by the astronauts of Apollo XIV—a lasting tribute to the uniquely enduring properties of cobalt-base alloys.

Substitution Efforts

Hastelloy alloy X, which has become Haynes International's most successful superalloy, was the product of a cobalt substitution program. In the early 1950s, the nation was faced with critical shortages of a number of metals. Each of these, including cobalt, chromium, tungsten and columbium was assigned a "strategic index number." Multimet alloy and Hastelloy alloy C, the major wrought high-temperature alloys of their day, had very high strategic indices. Company engineers worked diligently to develop an alloy that would satisfy the engine builders without the use of scarce metals. The result was Hastelloy alloy X, a Ni-22Cr-19Fe-9Mo alloy which has been a major product at Haynes ever since.

By coincidence, several months after the development of alloy X, a leading candidate alloy being tested for combustors at the Pratt & Whitney laboratories in Middleton, Connecticut failed. Alloy X was tested in its place and was ultimately selected for use in the JT3D engine for the eminently successful Boeing 707 jet. The rest is history. Today, virtually every gas turbine engine in the Western world has combustion zone components made of alloy X, if not its successor—Haynes alloy 188—or, more recently, Haynes alloy 230.

The metal supply situation was more relaxed when alloy 188 was developed in the late 1960s. The main objective was to combine or surpass the best properties of the two leading cobalt- and nickel-base sheet alloys. The new alloy was to have equal or better oxidation resistance than alloy X and better strength qualities than alloy 25. In addition, the new alloy was to show better post-aging ductility than either alloy. The result was a cobalt-base alloy (Co-22Cr-22Ni-14W-02La). Alloy 188 has taken its place in the more critical service environments while alloy X still remains the workhorse material of the industry.

Metallurgists were thinking more about the effects of an insurrection in Zaire when the most recent combustor alloy was developed in the early 1980s. This is Haynes alloy 230 (Ni- 22Cr-14W-2Mo-.02La), which has no intentionally added cobalt and relies more on tungsten than molybdenum for its outstanding high-temperature strength. Alloy 230 possesses the best combination of strength, fabricability, oxidation resistance, and thermal stability of the three materials. It is under serious consideration for use in advanced turbine engine designs by many engine manufacturers.

Necessity is indeed the mother of invention. The same forces that drove the development program that produced alloy 230 also gave rise to efforts directed at reducing cobalt in "gamma prime" alloys. One that was investigated was Haynes alloy 263, originally developed by Rolls Royce, Ltd. as a sheet metal combustor alloy. The alloy's original composition called for 20 percent cobalt, but tests showed that it could be made with half of that amount with little degradation in properties. This, in effect, would double the amount of cobalt available for that alloy. Similar programs were carried out with the cobalt-base, wear-resistant alloys, eliciting the consensus that effective, wear-resistant alloys can be produced with less cobalt.

Chromium, however, still remains a substantial challenge. Nickel aluminides and silicides are being evaluated, as are other material systems, but the prospects for viable commercial materials of note are not very good in the short run.

Manganese

In steel production, manganese is used mainly as a refining agent because of its great affinity for sulfur, a harmful impurity. Less than 15 tons of manganese are used annually at Haynes. Advanced melting and refining practices, such as argon-oxygen-decarburization and electroslag remelting, have reduced the need for manganese in alloy making.

Vanadium

Vanadium was once used extensively as a grain refiner in the production of high performance alloy castings, mainly the corrosion-resistant Ni-Cr-Mo alloys. When Haynes terminated the primary foundry casting business in 1974, the need for vanadium was virtually eliminated. However, it was found that small amounts of vanadium had a positive effect upon aqueous corrosion resistance. The metal was retained, in amounts less than 0.35 percent, in alloys such as Hastelloy alloy C-22 and Hastelloy alloy C-276.

Titanium

Haynes has limited production of titanium mill products in the alloy Ti-3A1-2.5V (containing 2.5 percent vanadium, which significantly increases strength). Two percent is the solubility limit for vanadium in titanium. The remaining one-half percent promotes the formation of a second phase that increases the alloy's usefulness by making it amenable to heat-treatment. Ti-3A1-2.5V has an unexcelled strength-to-weight ratio and is widely used for hydraulic tubing in both airframes and aircraft engines. A shortage in the supply of vanadium would be an inconvenience, but inferior materials could probably be used as substitutes at some expense in aircraft efficiency. Since the major-tonnage titanium alloy in the United States has twice the alloy content, shortages of Ti-3A1-2.5V alloy would be relatively insignificant by comparison.

Platinum

Platinum has very limited use in high performance alloys. Several decades ago a slight decrease in the corrosion resistance of high performance nickel-base alloys was attributed to the increased efficiency of producers in extracting this and other platinum group metals from nickel ores, but this was only conjecture. If platinum, manganese, and vanadium all were to disappear from the periodic table, as well as South Africa, human ingenuity being what it is—there would still be high-performance alloys.

Recycling

Recycling is an economic necessity and has long been a way of life in the production of high-performance alloys. Typically, half of every new heat of alloy melted at Kokomo consists of process scrap. Great care is taken to collect and segregate by alloy-grade, shearings, croppings, grindings, and swarf. Even slag from melt furnaces is milled to recover metallic values. Until recently, melt-shop smoke was collected in bag filters and recycled. But today's low metal prices make this uneconomical, and smoke solids are now used for landfill. Should metal prices rise significantly, smoke solids probably will be treated and go back into the electric furnace.

About 15 percent of each furnace charge at Haynes consists of purchased scrap. A concerted effort is made to obtain and use as much purchased scrap as possible. There are six highly qualified scrap dealers in the United States who deal primarily in Haynes alloys 188, 25, 718, R-41 and Waspaloy alloy. Haynes buys from these firms, usually in 20-thousand pounds lots to a guaranteed chemistry. Tungsten carbide tool inserts are purchased for tungsten values and pure molybdenum metal scrap is also sought.

Summary

1. With regard to South African materials, cobalt is very important to Haynes International, Inc., chromium is vital.
2. Steps must be taken to either stockpile adequate backlogs of chromium or to locate Western Hemisphere supplies.
3. For the present, commercial considerations must override strategic ones in the procurement of raw materials. Raw materials are the most important cost factor in the production of high performance alloys. Haynes International must remain competitive or perish.
4. A continuing search must be made for viable low-cobalt and low-chromium alloys.
5. Haynes feels that its recycling efforts are sufficient. Economics will govern the extent of recycling.

Art M. Edwards went to work for Union Carbide in 1948. Most of his career has been with Haynes Stellite Company, which was a Division of Union Carbide until 1970 when it became a part of Cabot Corporation, now an independent corporation, Haynes International, Inc. This is Mr. Edward's 40th year working with and writing about alloys.

South Africa's Minerals: How Big a Threat?

Hans H. Landsberg and Kent A. Price

Note. The following article by Hans H. Landsberg and Kent A. Price of Resources for the Future examines the risks to the United States of a cutoff in the supply of strategic minerals from South Africa. They conclude that although sanctions or embargoes could temporarily disrupt supplies, a complete cessation would be economically unaffordable to any regime in power in South Africa. Similarly, they conclude that the USSR does not have the economic or political wherewithal to preempt US access to South Africa's minerals. Although published in the summer of 1986, the authors' conclusions are still valid today.

In the article, Landsberg and Price also provide a thumbnail sketch of the advent of advanced materials and their potential to replace many of today's traditional strategic materials.

This article was first published in *Resources*, Summer 1986.

Riots, bombings, killings, military excursions beyond national frontiers, sharpened political factionalism, economic distress: the news from South Africa signals at least a new stage in that country's unhappy recent history and perhaps even an inexorable slide into bloody anarchy.

Rightly or wrongly, many of the most outspoken opponents of the white minority government of South Africa identify the United States as a key prop of the regime. Were the government to fall, and most observers think that would occur only after the most violent confrontations, the United States might well be the target of considerable resentment on the part of its successors, whatever their political orientation. What would this mean for what some consider the central US interest in maintaining effective relations with South Africa—the minerals trade?

Nature seems to have taken perverse delight in locating commercially and militarily essential minerals in places where their development and export fall far short of certain. Just as the vast majority of the world's known reserves of oil are in the perennially volatile Middle East, so nature has endowed southern Africa in general and South Africa in particular with rich deposits of nonfuel minerals. Of special interest in the what-if-South-Africa-falls question are four metallic minerals that are critical to one or another military application—manganese, chromium, cobalt, and the platinum group.

The United States depends overwhelmingly on imports for these four minerals. Imports account for nearly 100 percent of US consumption of manganese, and the import-dependence figures for chromium, cobalt, and the platinum group hover around the 90 percent mark. Although other sources for these minerals exist, some offer no political-strategic advantage—the Soviet Union, for example, exports both chromium and manganese—while others cannot match the cost and price structure made possible by the extensive deposits in southern Africa. Whatever the reason, the United States depends heavily on South Africa and its neighbors for these important minerals. And whatever one's views regarding the government of South Africa, it would seem that the incendiary situation there bodes ill for uninterrupted supply to the United States of materials critical to its national security.

But appearances may well be deceiving. What, after all, are the supposed threats to US interests from a movement toward black majority government in South Africa? Two plausible answers come to mind, one long-term and the other of more immediate consequence.

It is difficult to envision a scenario for South African rule that does not include substantial black participation and perhaps dominant political control. If and when a black government takes over, would it not be inclined to impose a minerals embargo on the United States to punish it for perceived support of the white regime? Or, short of a full black takeover, are not South Africa's waters so roiled as to make for easy fishing by the Soviet Union, either to control South African minerals or to deny them to the West? We think the answers to both questions are mostly negative.

Sanctions Ineffective

Examples of boycotts and embargoes intended to pressure foreign governments are easy to find. For many years, imports of chromium from Rhodesia (now Zimbabwe) were proscribed under United Nations sanctions in retaliation for that country's unilateral declaration of independence from the United Kingdom and for its racial policies. In 1973 the Arab oil-exporting countries embargoed exports to the Netherlands and the United States to protest the Middle East policies of these two countries and, in particular, US support for Israel. The United States itself for a long time prohibited trade in minerals and other commodities with mainland China and still imposes such sanctions on trade with Cuba.

But examples of *successful* national economic actions are rare, at least when imposed for political reasons. The Ian Smith regime in Rhodesia eventually gave way to black rule, but the UN sanctions had little or no effect on the outcome. The Netherlands and the United States obtained oil through third parties, thus negating the Arab oil embargo. And US trade sanctions applied to China and Cuba, although annoying, have not done much to damage their economies or to sway them from communism. Indeed, a case can be made that the resumption of US trade with China has done far more than the prohibition of trade to change its political and economic course.

In addition, the long-run costs of embargoes often are high for the imposing countries. Shortly after World War II, for example, the Soviet Union embargoed manganese shipments to the United States and other Western countries. This created considerable concern, because the Soviet Union was a major world supplier of manganese. The result, however, was not the intended change in US policies but rather the development of new manganese mines in India and elsewhere and the loss of Soviet world markets. Similarly, US embargoes of grain shipments to the USSR in 1980 merely shifted Soviet purchases to alternative suppliers and hurt US farmers.

Despite this poor record, embargoes continue to be attractive because they provide a visible means of expressing disapproval. When the use of stronger measures is ruled out, they give the appearance of bold action, whether or not they actually inflict any hardship on the offending country. So, for appearances, governments are likely to continue to impose embargoes from time to time and, in the process, disrupt the flow of mineral commodities in world trade. But the key word is *disrupt* not *stop*. Regardless of where they fall within the political spectrum, countries typically do not choose to forgo for extended periods the foreign exchange earned by an important mineral.

"Resource War"

Some hard-line cold warriors believe they see evidence of a "resource war" with the Soviet Union that threatens the security of mineral supplies from abroad, and particularly from southern Africa. In its most pointed version this supposed conflict is seen as part of a comprehensive plan to deny the West access to the mineral wealth of southern Africa, thereby complementing a Soviet desire to control the oil resources of the Middle East, in which the invasion of Afghanistan was a preliminary step. This jittery view of geopolitics is reinforced by highly visible acts, such as the unfurling of the Soviet flag at funerals of black militants killed in South Africa.

What is not clear in these speculations is how the Soviet Union could pursue a resource war (even assuming it actually wanted to, the evidence on that point being hardly convincing). Clearly the country does not have the foreign exchange to outbid Western consumers for mineral supplies from southern Africa. Nor does it seem likely that the Soviet Union is prepared to commit large numbers of its ground forces to the region, given the logistical difficulties and the likely political repercussions. If it were prepared to confront the West so openly, it would find the Middle East an easier and more attractive target.

This leaves the possibility that the Soviet Union would exploit political and tribal conflicts and support indigenous opposition to South Africa. But while such a strategy can be pursued with much less risk and cost and might extend Soviet influence in the area, it is unlikely to lead to actual control. With fresh memories of the past, neither a new South Africa nor the other nations of Africa will meekly surrender their hard-won independence to a new foreign power. Moreover, their own national interests, not those of the Soviet Union or world socialism, are likely to receive highest priority. Indeed, Soviet pressures to embargo mineral exports to the West probably would backfire, for such exports correctly are perceived as a vital source of foreign exchange and economic development. When new African Marxist governments have come to power—even with the help of the Soviet Union, as in Angola, Guinea, Mozambique, and Zimbabwe—they have tried to encourage, not cut, their mineral exports to the West and have strengthened other economic ties as well.

For these reasons, the resource war thesis appears implausible. Imports of manganese, cobalt, chromium, platinum, and other mineral commodities from southern Africa from time to time may be interrupted. The region still is going through a turbulent transition as it sheds its colonial past and adopts black majority rule. And the odds on wars, embargoes, rebellions, and even cartel attempts are far from trivial as newly independent states grapple with internal difficulties and hostile neighbors. But to attribute the insecurity of mineral supplies from this region to Soviet policies in pursuing a resource war misses the fundamental causes of this insecurity. Indeed, the notion already has lost much of its visibility and appeal.

New Frontier

Perhaps ironically, some years hence we may wonder why all the fuss was made about the security of conventional minerals. Visible on the horizon are what have been labeled *advanced materials*. The list is long and promising—materials derived from hydrocarbons, grossly called the petrochemicals; the graphites, more specifically graphite fibers that are hydrocarbon-associated, both in their natural and synthetic forms; materials of nonfuel mineral origin, like germanium, silicon, zirconium, and gallium, used either alone or in a vast variety of mixtures, including ceramic powders; and the so-called composites that draw on two or more of the new materials plus one or more of the conventional ones. And there is the use of new ingredients in old-line metals. For example, a new aluminum-lithium alloy, soon to be available commercially, is likely to become a significant structural metal, especially in the aircraft industry. The diversity of potential uses for the new materials is enormous. This is especially true of new applications in telecommunications, in transportation (both conventional and novel, such as in aerospace), in power generation, and in the electronics field generally.

Moreover, it is all but certain that the developing countries will not relive the materials history of Western Europe, the United States, or Japan. They will not go through a steel, copper, lead, and zinc stage but rather will employ the newer materials, above all the hydrocarbon-based ones, like plastics, even in uses that traditionally have employed the major metals. Indeed, many of them even may skip building the once-conventional basic infrastructure of railroads, ocean-going passenger and freight ships and tankers, large commercial buildings, bridges, and perhaps much heavy industrial machinery. No one can foretell the shape of the future, but it seems safe to predict that it will resemble the past only faintly.

But the advanced materials face a great many hurdles, and progress toward common use may not be swift. High cost is only the first of a host of important questions about the length of the commercialization process, the location and magnitude of eventual effects, and, by indirection, the consequences for conventional materials. The new materials frontier will not be crossed next month or next year.

In short, those who worry about the possible loss of critical and strategic minerals from South Africa and perhaps its black-ruled neighbors are not preoccupied with yesterday's problems or with needs soon to be obsolete. We may stand at the edge of a new age of materials, but for now the needs are real and Americans ignore the tinderbox in South Africa only at some economic as well as political peril. Our point is not that the minerals in question are unimportant or that cutoffs in supplies are impossible. Rather, we believe that South Africa's exports of minerals to the United States are not so precarious as to be the dominant factor in US relations with that country. Leaders in the United States should act in what they perceive to be the best interests of this country and of the people of South Africa. This is difficult enough without a misleading image of US dependence on South African minerals.

Interestingly, our view in effect is supported by the Reagan administration's July 1985 proposal to revise the US strategic materials stockpile. Simplified, the proposal would delete platinum from the stockpile, cut the cobalt and chromium goals by at least 75 percent, and reduce manganese to half its current goal. Whether intended or not, this proposal yet to be acted upon by Congress surely implies a relaxed attitude toward future dependence on South Africa's minerals—an implication to be taken all the more seriously as it comes from an administration that hardly can be charged with inattention to national security.

Kent A. Price is Assistant to the Chancellor of the University of Maine System. He is responsible for University relations with the state legislature and the news media, for system publications, and for public affairs.

Hans H. Landsberg is an economist and Senior Fellow Emeritus and Consultant-in-Residence with Resources for the Future.

Section II. The Public Sector

Why We Need a Strong National Defense Stockpile

Charles E. Bennett

Note. In the following article, Congressman Charles E. Bennett, Chairman of the Subcommittee on Seapower and Strategic and Critical Materials of the House Committee on Armed Services, stresses the need for a strong National Defense Stockpile. Congressman Bennett makes an analogy between the stockpile and an insurance policy against risks to the nation's security. According to the Congressman, the National Defense Stockpile is, in effect, an inexpensive guarantee against unforeseen disruptions in US supplies of strategic minerals, especially during time of war. The Congressman also stresses the need to keep the stockpile apolitical, as well as the need to make it dynamic—modernizing it and strengthening it as the country's military needs change.

In 1986 the United States imported raw and processed materials valued at $39 billion ($35 billion of which was in processed form) from foreign sources. Since the late 1950s the United States has become increasingly dependent on foreign sources for its supplies of minerals and energy. By 1980 US dependence on raw materials had increased to 25 percent of consumption, and we were more than 50 percent dependent on foreign sources for some 25 strategic and critical materials. Currently we are heavily import-dependent for such materials as bauxite and alumina (97 percent), cobalt (92 percent), columbium (100 percent), industrial diamonds (92 percent), manganese (100 percent), nickel (78 percent), platinum-group metals (98 percent), tantalum (91 percent), etc.

The strategic and critical materials are essential to the production of Army, Navy and Air Force armaments; aerospace equipment and jet engines; radio and electronic equipment, and medical equipment. In addition, these materials are necessary for the production of steel, including special purpose alloys, specialty steels, and stainless steel; petroleum refining equipment; oil and gas pipelines; cutting tools; rock drilling bits; etc.

George Santayana, in the *Life of Reason*, said, "Those who can not remember the past are condemned to repeat it." I'm afraid that we often ignore the lessons of history with respect to the National Defense Stockpile. It can be said with certainty that national leaders such as Truman and Eisenhower, who witnessed firsthand the enormous difficulties that were caused by not having sufficient strategic and critical materials stockpiled and on hand at the beginning of World War II and Korean war, were advocates for a strong national defense stockpile.

Policy-makers today should study the serious and costly bottlenecks that occurred in these previous conflicts and be reminded that the vast bulk of our military equipment and munitions were produced after the conflict began.

At the end of World War II the United States possessed unquestioned air superiority, naval superiority, and nuclear superiority. Our sources of mineral supply in Africa were under the control of our allies—Britain, France, Belgium, Spain, and Portugal. In addition, most of the industries in Latin America were run primarily by American companies.

All of this has changed. In addition, events in South Africa remind us that the United States is heavily dependent on that nation for our supplies of chromium, platinum, manganese, and vanadium which are essential to our steel, aerospace, and electronics industries.

What follows is an attempt to provide a brief history and purpose of the stockpile, to demonstrate that the stockpile is a cost-effective insurance policy in the event of a national emergency, to review the current status of the National Defense Stockpile, and to propose some suggestions and recommendations as to how the stockpile can be restructured and strengthened within the context of current national budget priorities and constraints.

History and Purpose of the Stockpile

Since the enactment of the Strategic and Critical Materials Stock Piling Act of 7 June 1939, the policy of the US Government has been to maintain a strong national defense stockpile. This policy has enjoyed the bipartisan support of the Congress. The purpose of the stockpile is to provide materials to supply the military, industrial and essential civilian needs for national defense purposes and to preclude a dangerous and costly dependence by the United States on foreign sources for supplies of such materials in times of national emergency.

Current National Defense Stockpile policy originated out of the experiences of World War I, World War II, and the Korean conflict. For example, in World War I, materials prices escalated dramatically, and the US industrial base was unable to mobilize quickly to provide the supplies and equipment needed to equip the American Expeditionary Forces. Notwithstanding the tremendous outlay of funds, the United States fought World War I, in large part, with guns, munitions, airplanes, and other materiel borrowed or purchased from France and Great Britain. Between World War I and World War II, numerous studies and recommendations called for the stockpiling of strategic materials to prevent the mistakes that occurred during World War I. However, the United States entered World War II without an adequate stockpile of strategic materials and thereby lost the advantage of bringing the nation's superior industrial capability fully to bear at the outset of hostilities. This failure to prepare in advance required dramatic decisions: to allocate the limited supplies available, to restrict civilian consumption, to require the use of substitutes, to make technological changes in manufacturing processes, to expend vast sums to develop low-grade domestic sources of supply and, above all, to use shipping badly needed for military purposes. Warnings of these consequences were sounded prior to the war but to no avail.

At the close of World War II, the Stock Piling Act of 1946 was enacted to ensure that an adequate National Defense Stockpile would be available for any future emergency. However, the United States entered the Korean conflict with a woefully inadequate stockpile. As a result, the frantic effort to acquire strategic materials completely disrupted world mineral prices. As an example, between May 1950 and January 1951, in a period of little more than six months, the price of tin increased 135 percent, antimony 71 percent, lead and zinc 45 percent, and crude rubber 157 percent. Similarly, substantial increases in prices occurred in copper, tungsten and numerous other vital materials, which increased the cost of the stockpiling and rearmament programs.

As a result of these experiences, the National Defense Stockpile became an integral part of US national security policy. However, since the early 1950s, stockpile policy and goals have fluctuated dramatically as a result of changes in policy-makers' stated perceptions about the strategic threat to the national security such as the location, duration, and intensity of the wartime scenario. Presidents Truman and Eisenhower were strong stockpile supporters because of their experiences during World War II and Korea, and stockpile goals were initially based on supporting a five-year national emergency similar to World War II. However, as war memories dimmed in the early 1960s, President Kennedy expressed the view that the stockpiles were too large, and goals were reduced. Large amounts of strategic and critical materials were classified as surplus to requirements.

During the Johnson administration large quantities of copper, aluminum, nickel, etc., were sold to hold down material prices which began to rise due to the demand created by the Vietnam war.

On 16 April 1973, President Nixon dramatically reduced stockpile goals by concluding that the stockpile was only necessary to provide for a one-year emergency period on the basis that (1) the world was more peaceful than during the preceding years of the stockpile, (2) improved technology would make it easier to find substitutes for scarce materials, and (3) if a war did occur and last longer than one year, the one-year period would be sufficient time in which to mobilize and sustain the nation's defensive effort for as long as necessary. Furthermore, it was argued that any conflict would be a nuclear war, swift and decisive, and any protracted conventional conflict of relevant severity was unlikely. As a result, billions in stockpile sales ($21 billion in FY 74 alone) were used by the Nixon administration to reduce budget deficits and to hold down materials prices as an anti-inflation device. However, on 5 March 1975, the Subcommittee on Seapower and Strategic and Critical Materials of the House Committee on Seapower and Strategic and Critical Materials of the House Committee on Armed Services rejected President Nixon's one-year objective and approved the requirement for a three-year stockpile objective.

Both the Ford and Carter administrations approved stockpile goals based on a three-year national emergency following their own exhaustive stockpile policy reviews. In an attempt to promote some stability to the stockpile program, the Congress passed the Strategic and Critical Materials Stock Piling Act of 1979 (PL 96-41) which established the National Defense Stockpile Transaction Fund. This fund was established to receive receipts from the sale of excess or obsolete stockpile materials which were to be used to purchase more critically needed materials. In addition, the Congress made it clear that "the purpose of the stockpile is to serve the interest of national defense *only* and is not to be used for economic or budgetary purposes." Upon assuming office, President Reagan instituted the first significant stockpile purchases in more than 20 years with the acquisition of cobalt and bauxite, followed by purchases of beryllium, iridium, nickel, palladium, quinidine, rubber, tantalum, titanium and vanadium, totaling over $367 million. Later, in his 5 April 1982 Report to the Congress, he reaffirmed his policy of reliance on the stockpile to meet military, industrial, and essential civilian needs in support of the national defense in time of national emergency.

This policy was short lived, however, when Budget Director David Stockman concluded in 1983 that stockpile goals could be reduced by $10 to $13 billion. This pronouncement led to a further National Security Council stockpile review.

National Security Insurance

Former President Eisenhower stated in 1963, "The Nation's investment in these stockpiles is comparable to the investment made in any insurance policy. If an emergency does not arise, there are always those who can consider the investment a waste.

If, however, the investment had not been made and the emergency did arise, these same persons would bemoan, and properly so, the lack of foresight on the part of those charged with security of the United States. I firmly rejected the policy of too-little, too-late stockpiling."

Why do we need an insurance policy? Because there are significant risks to our national security that are beyond our ability to control or beyond our ability to forecast with any degree of certainty.

What are these risks?

1. Not having sufficient material on hand to support the immediate industrial surge requirements to produce the weapons of war or the infrastructure necessary for their production.
2. The strong possibility that the United States will be unable to secure adequate supplies from foreign sources for those materials that cannot be produced in sufficient quantities in the United States.
3. The possibility that domestic production may not be able to provide the necessary quantities of materials or to produce them in a timely manner.
4. US naval warships will have to be diverted to protect commercial shipping bringing raw materials from abroad. Consequently, naval assets may not be effectively deployed for critical combat operations.
5. The necessity to expend large federal outlays to encourage domestic materials production, to replace naval and commercial shipping losses to enemy action, and for the purchase of strategic and critical materials at sharply higher wartime prices.
6. The danger of diverting scarce equipment, materials, energy, transportation, and manpower from direct war production to produce strategic and critical materials in short supply.
7. The possible delays in obtaining materials from domestic or foreign sources.

8. And, most importantly, the danger of not having materials immediately on hand to equip our fighting men and women with the necessary quantities of war material to bring the war to a speedy conclusion. This last point is crucial since it could mean the unnecessary loss of additional lives should the war effort be delayed for lack of adequate materials.

Individuals purchase health insurance to protect themselves or their families from catastrophic illness which could financially ruin them. The National Defense Stockpile is an insurance policy that can provide significant national security protection at a small premium. The current stockpile was accumulated over many years yet its value represents less than 3 percent of the annual defense budget of approximately $300 billion. A stockpile of $10 billion of strategic and critical materials could be converted into finished goods representing $40 to $100 billion in needed war materials. The nation should not risk an inadequate stockpile because the risks and uncertainties involved are too great. Finally, the premiums for this insurance policy are invested in strategic and critical materials, which are assets owned by the government which do not deteriorate but generally rise in price over the long run. It is, I believe, a wise investment which provides a significant national security benefit.

Adequate Stockpile to Support a National Emergency

There can be no argument about the need for large quantities of stockpile materials in the event of a conventional conflict between the United States and the Soviet Union. However, there were those in the Reagan administration who argued strongly that we must continue to support the massive conventional military buildup, but at the same time recommended that most of the stockpile materials be sold because a conventional war was unlikely since it would rapidly escalate to nuclear war. The administration cannot have it both ways. We are either overspending for conventional defense or ignoring the importance of the stockpile to our national security.

I believe we should continue to provide funds for the buildup of our conventional forces while at the same time giving greater attention to improving and strengthening the National Defense Stockpile. If war does occur, it will probably be a conventional war, and if we are not able to win it, we'll probably be forced to escalate to a nuclear war. What a dull and dangerous plan! Particularly when we could plan to win it conventionally.

The greatest challenge our country has today is not the terrible deficit we have, it is not arms control; it is the need for the decision of the United States and our allies to be prepared to win a conventional war wherever it may occur, or at least prevent any decision to use nuclear weapons or delay such a decision for some appreciable length of time until negotiations can lead to a cessation of hostilities. That is the biggest mistake humanity is making in 1987, not being prepared to win a conventional war. Because by going straight for nuclear war, with there being no way to win a nuclear war, we would be planning for a disaster for all mankind.

The stockpile is an important component for any preparation for a conventional war emergency. Consequently, the basic question facing the Congress and the administration is the appropriate size of the stockpile. However, the determination of realistic national emergency requirements for strategic and critical materials is fraught with difficulties since the United States must be prepared to respond to a range of conflict scenarios and contingencies. According to the President's Materials Policy Commission in 1951,

> Defining the specific requirements for wartime materials would require a comprehensive study of alternate strategic plans including such factors as the geography and logistical difficulties of war wherever, whenever, and by whatever method it might be fought. These factors could be more easily estimated by an aggressor since the act of aggression itself determines the timing and location of attacks. For the free nations whose efforts are directed toward averting aggression, forecasting the variable particulars of war demand presents great difficulty.

If the United States is to win another war, the ability of the American defense industry to surge production of combat consumables is absolutely essential. The magnitude of the requirements for a conventional war is staggering. During World War II, the US defense industrial base produced 310,000 aircraft, 88,000 tanks, 10 battleships, 358 destroyers, 211 submarines, 27 aircraft carriers, 411,000 artillery tubes and howitzers, 12,500,000 rifles and carbines, and approximately 900,000 trucks and motorized weapons carriers. From 1944 to 1945 more than 40 percent of the total US output went for war purposes.

Discussing wartime demand and potential materials supply during any future war, the President's Materials Policy Commission observed,

> Once conversion to military and essential civilian production was accomplished, the "normal" wartime demand would have to be expected to continue for a considerable time. The last two wars lasted 4 years or more. It is conceivable that revolutionary weapons might shorten a war, but it would be unwise to gamble on that. The prudent theory, in terms of historical perspective, would be that war would continue with increasing violence for a number of years, and that the demand for materials would continue high and perhaps even rise as industrial capacity moved upward.
>
> The main problem of supply is that the United States' own resources base would not be adequate. The fact that our allies would be increasing their demands at the same time would limit the supplies from outside the country on which the United States might draw. Moreover, domestic facilities might be bombed. Free countries which supplied key materials might be subverted or invaded, or enemy action might damage facilities or sink cargoes.

The above discussion illustrates that it is unlikely that anyone can safely predict what the nation's requirements for materials might be for some future national emergency. Therefore, it could be dangerous to rely on any estimate which does not provide some safety factor or margin for error and is not geared to protect the nation against the worst-case scenarios.

It is my considered belief that the goals for the ultimate size of the stockpile should be based on the worst case war scenario even though we may decide that the cost is too great to be fully implemented. It should be noted that in a report to the House and Senate Armed Services Committees from Deputy Secretary of Defense Taft dated 17 March 1987, the Department of Defense stated "The estimated cost to build and maintain the planning force under the three-year total mobilization scenario is approximately $1 trillion above current five-year defense plan levels."

This Department of Defense estimate is in direct conflict with the recent National Security Council stockpile study which was based on the assumption of less than total mobilization and a war whose intensity diminished after the first year and was concluded in three years, rather than the first three years of a war of indefinite duration. What serious student of military strategy and the defense of Europe would publicly subscribe to the unlikely scenario put forth in the National Security Council study? It is certainly not what the existing statutory provision requires.

Policy-makers must be provided with an objective estimate of stockpile requirements based on a worst-case wartime scenario so that acceptable funding priorities can be established for restructuring the stockpile in a manner that will maximize our national security.

Current Status of the Stockpile

The current National Defense Stockpile is composed of strategic and critical materials largely transferred from the Reconstruction Finance Corporation after World War II, materials purchased during the Korean war, including Defense Production Act inventories, $16.0 billion of materials obtained by the US Department of Agriculture program, which bartered surplus agricultural commodities for these materials, and other purchases such as the recent Reagan administration acquisitions.

Over this same period, approximately $9 billion of materials have been sold from the stockpile, leaving about $8.3 billion (based on 29 September 1986 prices) of strategic materials stored in some 107 depots in the United States.

The following charts show an analysis of stockpile materials with excesses and shortages based upon criteria required by law and not reduced as suggested in the recent National Security Council stockpile study.

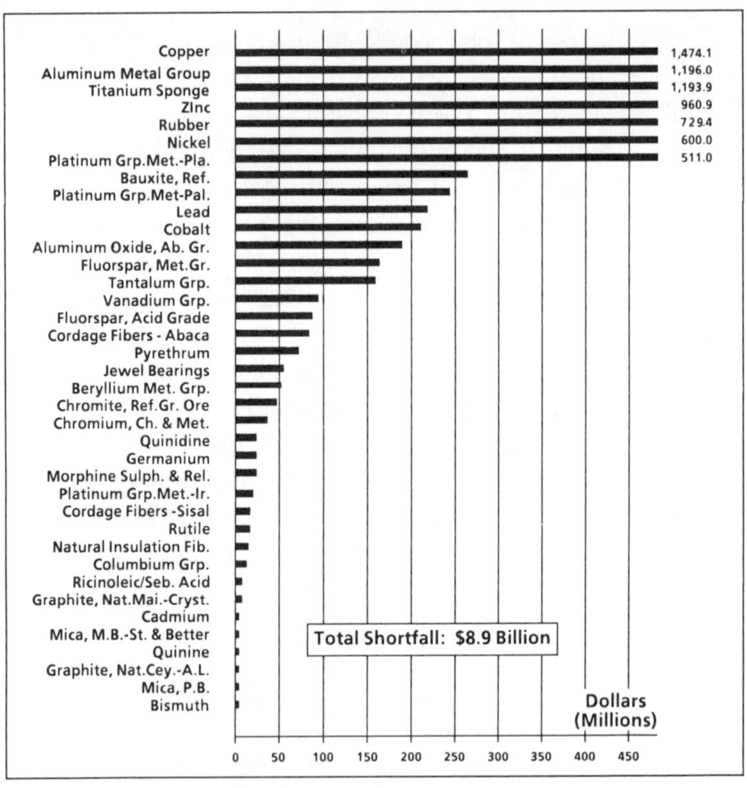

Shortfalls in Inventory of Stockpile Materials as of 30 September 1986.

Restructuring and Strengthening the Stockpile

If we are to satisfy the question "What is best for the United States?" we must restructure and strengthen the stockpile. Congress and the administration must agree on an objective analysis of the quantity and quality of each material that is to be included in the stockpile. This analysis must be made by the best qualified persons in the nation and should include panels of experts from the mining, milling, smelting, fabricating, and using segments, in conjunction with government representatives. This analysis should be considered a technical analysis and not a political, economic, or budgetary exercise. The analysis must be based on

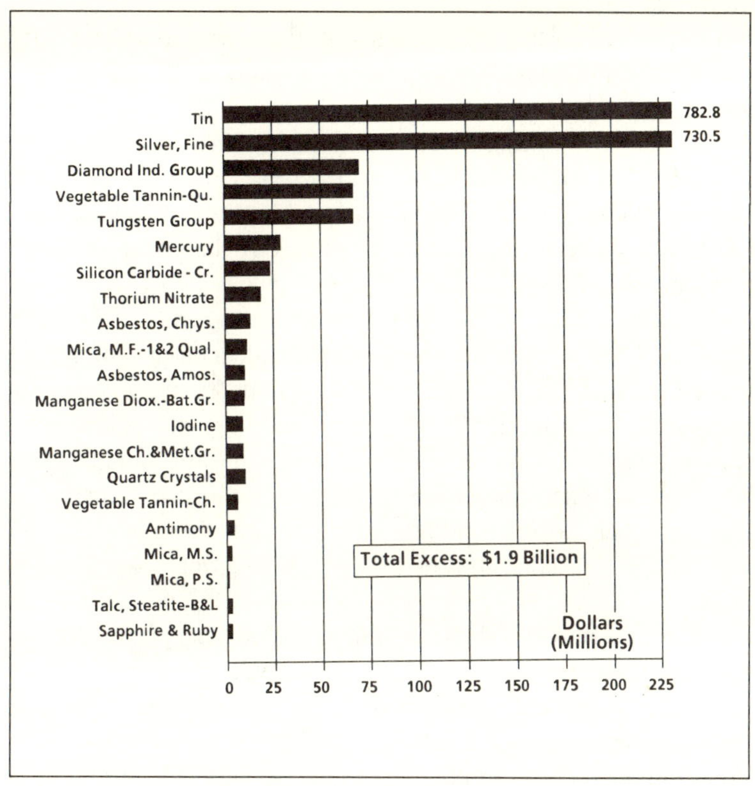

Excesses in Inventory of Stockpile Materials as of 30 September 1986.

a worst-case scenario that would include total mobilization of the economy for not less than three years of a war of indefinite duration. After the long-range requirement for each material has been established, it should only be changed for unusual circumstances. In this regard, a full-blown review of stockpile requirements should be made every five years so that new materials can be identified and incorporated, obsolete materials eliminated and sold, and identification of the form and specifications for stockpiled materials which should be modified.

When specifications and requirements have been established, then we must develop a long-range plan for acquisition, disposal, and upgrading so that the materials in the stockpile will be sufficient to be immediately responsive to wartime requirements. The form of each material must incorporate as much labor, energy, and transportation as possible before hostility begins.

We need to resurrect the original concept of the National Defense Stockpile Transaction Fund as being a true revolving fund and to put a stop to the continuing efforts by the Office of Management and Budget to sell off larger quantities of stockpiled materials and diverting the proceeds to the general fund of the Treasury. I am just as concerned about our federal budget deficit as the Office of Management and Budget, but I think it is foolish to liquidate our important stockpile assets to obtain funds simply to reduce the deficit. We could go a long way toward strengthening the National Defense Stockpile through the use of the Transaction Fund approach, without adding a single dollar to the federal deficit.

We need to constantly assess which materials from foreign sources would be most vulnerable to supply disruption in case of a war. We need to establish a planning process that prioritizes the acquisition and upgrading of needed materials where the greatest vulnerability and risk are indicated. For example, on 7 January 1987 the Department of State certified that quantities of materials being imported from South Africa such as antimony, chromium, cobalt, industrial diamonds, manganese, platinum, rutile, and vanadium were essential to the economy or to the defense of the United States and are not available from other reliable and secure suppliers and, therefore, should be exempted from the Anti-Apartheid Act of 1986 and its sanctions against South Africa.

We also need to have a process which will identify new materials requirements resulting from new technologies and new developments. I am thinking of such materials as the rare earth metals, selenium, tellurium, hafnium, etc., and I am certain that many other similar materials will become essential to our defense and, if unavailable in the United States, will need to be included in the National Defense Stockpile. At the same time, the process needs to identify and eliminate from the stockpile obsolete materials as well as materials so far below specifications it would be more economical to purchase new materials than to upgrade.

The US Government has unnecessarily classified essential information about the stockpile and, therefore, it has been unavailable to interested persons. Information about stockpile requirements and stockpile proposals for the acquisition, disposal, or upgrading of materials needs to be understood by a variety of persons. We need to bring this information out into the open so that informed opinions can be considered which will have an impact on decisions being made about the stockpile.

For the past 25 years various administrations, regardless of who was in the White House, have demonstrated that they are unwilling to place the more important national security considerations above narrow economic and budgetary considerations.

Consequently, I believe Congress must exercise its responsibility to authorize the quantity and quality of each material to be maintained in the National Defense Stockpile, just as it now authorizes the number and kinds of airplanes, tanks, artillery tubes, and warships.

Florida Congressman Charles E. Bennett is chairman of the Seapower Subcommittee of the House Armed Services Committee. He is also a member of the Procurement and Military Nuclear Systems Subcommittee; and is a member of the House Merchant Marine and Fisheries Committee and Merchant Marine and Coast Guard Subcommittees.

Surge Production, the Industrial Base, and Other Strategic Material Concerns within DOD

Richard E. Donnelly

Note. The following essay is based on an interview with Richard E. Donnelly, Director for Industrial Resources for the Department of Defense. During the interview, Mr. Donnelly stressed new DOD initiatives designed to stem the erosion in the military vendor industry in order to ensure an adequate surge production capability, i.e., to allow DOD's suppliers to quickly gear up production in a matter of only a few months in order to support the military in the event of a sudden, if short-lived, emergency. Mr. Donnelly also addressed the changing material requirements for today's more technically sophisticated weapons systems, as well as some of the major inadequacies of the current approach to national defense stockpiling.

The mission of the Office of Industrial Resources is to establish policies and procedures to ensure effective domestic industrial infrastructure available to support the US five-year defense program and to be able to surge or mobilize or otherwise increase production and support in the event of emergency situations. Industrial resources are defined as raw materials, semifinished parts, and components necessary to produce military equipment. These resources also include manufacturing technology, i.e., the capability to make things in a modern way, DOD-owned plants and equipment, as well as manufacturing equipment that DOD occasionally provides to privately owned contractors. Occasionally, DOD also is concerned about industrial manpower, and the availability of energy as a process fuel. To use a corporate analogy, the office is similar to the corporate office of a large conglomerate like General Motors. They have operating divisions such as Chevrolet, Cadillac, Buick, etc. The operating divisions of the Office of Industrial Resources are the Army, Navy, and Air Force. As such, our office tries to write policies and procedures for the military departments to operate by.

Decline in the US Industrial Base

The main concern we have with the industrial base over the past few years has been that at the prime contractor level, for example, aircraft assembly plants, there is generally available capacity. Some might say there is excess capacity and that may be the case in the aerospace area. But when you move away from the prime contractors down to vendors and suppliers, we find out that there is a lack of competition and a dearth of surge production capability. Many of these vendors and suppliers have been hurt by import penetration and have failed to modernize to the extent necessary for them to remain competitive. The extent that this vendor base continues to erode has implications for surge production in the event of an emergency.

Surge Production and Mobilization

Surge production is not designed for the classical full mobilization. Surge production is a way usually within existing brick and mortar, but not necessarily, for us to rapidly expand production of key items identified by the Commanders-in-Chief that will be needed in the early stages of some type of emergency. A surge capability functions as a deterrent to war but should the deterrent fail, it provides the capability to rapidly increase defense-related production in order to end a war more quickly. Failing that, surge production certainly provides a stronger footing to be able to move toward full mobilization should that be necessary.

The concept of surge production creates a capability to double or triple production of an existing weapons system within six months of a decision to do so. Basically, components and materials with long lead times to produce are acquired and kept in a "rolling inventory" by the contractor to enable a boost in production of weapons or equipment in a hurry. In some cases special tooling and test equipment would be necessary to meet an increased production rate. The TOW II missile, which is a tank killer, provides a good example. Without a built-in surge capability, it normally would take on the order of 12 to 18 months to double production, and that higher rate could not be sustained indefinitely because the pipeline of the component parts incorporated in the TOW II would soon dry up. Once that happens one cannot fill it back up again. By establishing a rolling inventory in the plant for long lead-time items like the gyro and the beacon assembly, the contractor could draw from that inventory while at the same time placing orders to keep a sustained inventory. In sum, we can now go from a production rate of 1,800 missiles per month to 3,600 missiles per month in less than 6 months and sustain that indefinitely. Without a surge this could not be sustained.

Such investment in long lead-time items is in a way almost "free." In the event we do not go to war and, thus, do not require a surge in production, those long lead-time items would be used anyway as the production buy is completed. In effect, we just buy them earlier. Although there are additional costs for special tooling and test equipment under this concept, these costs are necessary.

If we do not go to war, critics can say that we bought one more piece of tooling that we did not need. Although these critics argue that most plants operate on one shift per day and that by going to three shifts a day one could quickly triple production, that argument falls apart rapidly. Because special tooling and test equipment are so very expensive, they are usually run almost 24 hours a day, i.e., there is no residual capacity. With the policies we are now putting in place for new weapon systems, bottlenecks such as the above will be virtually eliminated.

Response to Emergency Situations

Although it is difficult to imagine a highly intense, long duration conventional war, we do plan for it. But from an availability of industrial resources standpoint, our main concern is the possibility of on-again, off-again "serial" emergencies such as localized wars—maybe involving our allies where we would be assisting them in some way, either directly or indirectly. However, under such conditions we could expect literally to be cut off from supplies of raw materials overseas for political or military reasons and for an extended period of time. Hence, there is an interest on our part in the availability of materials and stockpiling. By careful planning on our part, we hope to avoid production bottlenecks in certain key industries including castings, extrusions, electronic spare parts, optics, bearings, etc.—the "horseshoe-nail" kind of industries.

Technology Implications of Dependency

In the classic question of mobilization, semiconductors serve as a useful example. One could say from a purely industrial mobilization standpoint, perhaps we should not be so concerned about a cutoff from overseas supplies of semiconductors because they are not the type of raw material that has to be put into a break bulk carrier and shipped around the horn of Africa. Enough can be flown over in a couple of suitcases to ensure adequate supplies for months and months. Our concern about semiconductors is more of a technology-based question, a more basic type of foreign dependency. In the case of semiconductors, if foreign sourcing goes on long enough, there is the possibility that we will become technology dependent on overseas sources, that we will then be put in a position of having our trading partners provide us only second-best technology. Such a situation has infrastructure implications because we depend on a qualitative advantage for the electronics which go into the weapons and equipment we produce. Indeed, semiconductors are clearly essential for information processing, supercomputers, etc., that form the basis for our edge in weapons system design.

Initiatives in Government-Owned Industrial Facilities

The Department of Defense currently has about 67 government-owned, government-operated or government-owned, contractor-operated plants. Most of the government-owned facilities are ammunition plants, but DOD also owns some aircraft plants and other related facilities. We are attempting, where it makes sense, to get out of government ownership of plants and equipment. The Defense Department is a very poor landlord. It tends to put off modernization of its plants because of other budget pressures such as the purchase of weapons and equipment. Even when there is general agreement that it would be a good idea to go fix the roof at Plant so-and-so, the tendency has been to wait one more year in order to buy another F-15 fighter or Mark 82 bomb. As a result, DOD gradually has been divesting its ownership of such plants.

This program has been fairly successful. DOD recently sold three jet engine production facilities to the operating contractor, as well as two heavy press plants—one in Cleveland and one in Massachusetts. This program pays dividends in that we have found generally that as the ownership changes, then more modernization begins to occur. A privately owned plant needs to be kept modern in order to stay competitive. For example, regarding the heavy press plants in Cleveland and in Massachusetts, the contractors already have made investments in ancillary equipment to help make those presses more productive. So the change of ownership is done under a concept that says we do not need to own the plant; we just have to make sure that it stays there, that it is modern, and that it can produce the things that we need. Under this concept, formerly DOD-owned plants are now being maintained in a very productive, modern way—introducing, for example, flexible manufacturing systems that allow these plants to produce things better, cheaper, faster and so forth.

There is another area of concern, however. The above discussion refers primarily to plants acquired many years ago. According to the Defense Industrial Reserve Act and other related laws, the private industrial infrastructure must, by and large, be used to manufacture defense goods. Any investments that we do make in industrial facilities—defined as plants, industrial plant equipment, special tooling, special testing equipment, etc.—can only be done when we are convinced that industry is unwilling or unable to satisfy our needs with its own assets. We have to be careful, however, not to rely on private industry completely to come up with all the upfront investments in tooling and equipment. In some cases it may be necessary for DOD to put up seed money to make sure we have the right infrastructure in a fast changing technical environment. This policy is currently under review to see whether any additional changes need to be made.

Change the Stockpile?

DOD's difficulty with the stockpiling concept relates to changes in warfare concepts, defense policies, emergency scenarios, etc. From our standpoint, we have to be prepared to punch up production very rapidly if necessary, and the current stockpile does not allow us to do so adequately. Most of the materials in the stockpile were acquired back in the 1950s and many of them have not, naturally, kept pace with changing technology. In some cases, the wrong materials are in the stockpile inventory. Worse than that is the form of the materials that are held. Does it make sense to us to stockpile raw materials that could not have an impact under wartime conditions for 12 to 18 months? Our primary concern is do we have the right amount of manufacturing capacity and the necessary amount of materials in the right form to make a difference in the first several weeks and months of a war? A third concern is the lack of new high-tech materials in the stockpile. Indeed, under almost any imaginable emergency scenario where stockpiled material could really be useful, the current stockpile is basically inadequate. For example, it largely contains basic ores that would require a substantial amount of processing before they would be useful in supporting a war effort.

We also believe that there are other alternatives to stockpiling that need to be more carefully considered. Take high-purity silicon as an example. Right now the super high-purity silicon that is needed for many weapons systems and equipment comes from only one source, which happens to be 30 miles from the Czechoslovakian border. One might say, well, there is a good candidate for stockpiling. Why not stockpile in the United States the amount of high-purity silicon necessary to meet our emergency needs, whatever we think they are. If we get cut off, we can draw from the stockpile. Although that might be one way of solving this particular material need, maybe it is not the most cost-effective way. Perhaps it would be better to create a set of economic incentives that would stimulate US investment in manufacturing facilities for high-purity silicon. In fact, that is what we are now doing under Title III of the Defense Production Act. Projects such as this create incentives in the marketplace to initiate or expand capacity and supply. In the case of high-purity silicon, we are going to the private sector with a "purchase guarantee" that we will purchase X pounds of silicon per year. In effect, we are creating a competitive environment among several producers that assures them of a market while ensuring that we will have adequate supplies. DOD and the taxpayers come out ahead because we do not spend a dime until we get the high-purity silicon that meets the specifications. Industry undertakes little risk because it can go to banks for loans to put in the necessary manufacturing equipment, knowing that they will be able to sell what they produce. This creates a win-win situation.

As for the future of the National Defense Stockpile, it will be under review for some time to come. From a Defense Department perspective, we are pressing for a closer look at what kinds of new materials might be candidates for stockpiling. DOD is also looking carefully at the idea of stockpiling upgraded forms of materials instead of basic ores, i.e., materials and structures further up the processing train but not so far up that they become obsolete if they sit on the shelf too long.

Disruptions in Foreign Supplies

Despite the growing problems within South Africa, a major source of strategic minerals, there is less concern in DOD now than there was 10 years ago for a number of reasons. For one thing there is such a large amount of material in the stockpile today that even if it had to be drawn down in an emergency, it would be years before inventories would be depleted in any significant amount. For example, we are not overly concerned about cutoffs in the cobalt supply from Zaire and Zambia, since there is so much cobalt in the stockpile. Moreover, many of the new weapons and equipment that we are designing and producing today are not as materials intensive as others produced years ago. We simply do not need as much steel and ferroalloys as in the past when we had to construct a Liberty ship a week or whatever the rate was in WWII. Generally, we do not expect future emergencies to be particularly materials intensive. And the mix of materials that will be needed has also changed drastically. We are moving away from steel, copper, aluminum, and nickel toward composites made from graphite, carbon fibers, etc. Maybe these materials should be candidates for stockpiling or the new, high-reliability electronic materials.

We have always sourced southern Africa for chrome, cobalt, etc., and I think that will continue well into the future. But the nature of this dependency is changing, and that is some source of concern with DOD. The worry now is that fabricating capability is gradually moving to overseas markets. That is why it is important for DOD to maintain better visibility in the vendor and supplier structures that underlie our prime contracts. To reiterate, we believe the best way to deal with the problem is to stimulate US industry into becoming more productive. In other words, we need to fix the problem at the front end rather than trying to fix the problem at the rear through stockpiling or through restrictions on DOD procurement that say you have to buy from the United States.

Those are only "bandaids." The key is to find out why US industry cannot hold its own in the world marketplace, and it is not as simple as cheaper overseas labor. It is much more complex, and we are studying the problem within DOD to find out the root causes of these basic industrial shifts. The bottom line from our perspective is that the sun does not rise and set on the stockpile alone. That is why we are examining other factors that are eroding our industrial base.

Richard E. Donnelly is Director for Industrial Resources within the Office of the Assistant Secretary of Defense, Acquisition & Logistics. He is responsible for providing policy and planning direction within the Department of Defense (DoD) to ensure the readiness of a cost-effective industrial production base to meet peacetime and emergency requirements.

Mineral Economics in the Defense Community View

Richard Levine

Note. Richard Levine—one of the key players in the 1985 stockpile study when he served on the staff of the National Security Council—summarizes the background behind the study, its assumptions, methodology, and findings. The article originally appeared in the journal, *Materials and Society*, volume 10, number 2, 1986 (pages 217-19) and is reprinted here by permission.

The National Security Council (NSC) has been deeply involved with the nation's defense stockpile policy. In order to understand the NSC's involvement with the nation's stockpile policy it is important to understand the NSC.

The NSC is presided over by the President. In addition to the President, its statutory members include the Vice President and the Secretaries of State and Defense. The Director of Central Intelligence is the intelligence adviser to the council, and the Chairman of the Joint Chiefs of Staff is the statutory military adviser. The statutory function of the council is to advise the President with respect to the integration of domestic, foreign, and military policies relating to the national security.

An NSC meeting is intended to provide the President with an array of views concerning extremely complex issues so that balanced, well-reasoned decisions can be made. Over the years, a staff has been built up to serve the needs of the NSC. Today, the NSC staff, which numbers about 50 professionals, has the role of helping to prepare policy options for the NSC as well as the equally important job of serving as a "watch dog" in order to ensure that Presidential policy directives are carried out.

With this background in mind, the NSC staff's role in the formulation of the President's stockpile plan may be considered. With broad guidance from the Assistant to the President for National Security Affairs and several cabinet officials, the NSC staff directed a comprehensive review of the nation's wartime material requirements. In May of 1985, the NSC staff presented the results of its two-year study of the stockpile to the NSC. (This stockpile study involved over one hundred professionals from more than a dozen different agencies and departments.) At the May NSC meeting, 16 departments, agencies, and offices were represented so that the President could assess their views and benefit from their expertise. On the basis of this NSC meeting, the President adopted new stockpile goals that for most materials were substantially lower than the previous 1979 goals.

The President's endorsement of lower stockpile goals may have seemed paradoxical at first, given the administration's emphasis on defense preparedness. However, the new, lower stockpile goals are, in fact, more attuned to the nation's true defense requirements than the previous set of stockpile goals.

The Reagan stockpile goals were derived from a comparison of estimates of the nation's essential wartime material demands and reliable material supplies. If, for any material, the essential wartime demand was found to be greater than the available, reliable supply, a stockpile goal was established. The estimates of supply, demand, and the resulting stockpile goals were calculated by the following method. First, since by law the stockpile is to provide for the nation's essential material needs for a national emergency of at least three years of duration, a war scenario for stockpile planning was developed. The Joint Chiefs of Staff provided a war scenario that was consistent with the statutory requirements for stockpile planning as well as current defense planning guidance. This scenario, when contrasted with the 1979 stockpile planning scenario, defined a larger war with greater forces deployed over more battle fronts. The war scenario adopted for stockpile planning is extremely materials intensive since it involves a global war which is conventional and not nuclear in nature.

In order to estimate materials supply in a conflict, the Bureau of Mines did engineering estimates of the materials production and supply abilities of the world's material-producing nations including the United States. After these estimates were completed by dozens of professionals at the Bureau of Mines, the intelligence agencies of the US Government provided guidance concerning the wartime reliability of foreign supplying nations. The intelligence community, in making these recommendations, considered issues such as the potential for the disruption of these nations' internal transportation links by either combatant forces or terrorists. To complete these material supply estimates, the Department of the Navy and the Office of the Secretary of Defense estimated the wartime losses of ships carrying materials from reliable foreign producers to US ports. Thus, the calculation of material supply for the stockpile consisted of three related steps: (a) calculation of country-by- country materials production potential, (b) country-reliability assessments, and (c) shipping loss estimates.

In order to estimate US materials demand in the projected conflict, a projection of the United States wartime economy and defense establishment is required. In order to estimate the nature of the US economy in a war, the two most important factors that will shape the US economy in conflict must be considered. The first factor is energy supply. In World War II, the US was a net energy exporter. Although we are now not as dependent on foreign energy supplies as was the case 10 years ago, the US today—in peace and in a conflict—is a net energy importer. The restricted supply of oil in a worldwide conflict would have a substantial effect on the US economy. To aid in estimating this effect, oil prices were calculated in light of the adopted war scenario. The other major factor affecting the US economy in a conflict is increased defense expenditures. The Department of Defense estimated defense spending levels for the projected conflict. These spending levels are far greater than the current defense budget and demonstrably in excess of the defense spending estimates which formed the basis of the 1979 stockpile goals.

The projections of energy prices and wartime defense expenditures were used in a macroeconomic model to describe US GNP for the three war years. After a GNP growth path was developed, input-output models provided by the Federal Emergency Management Agency and the Department of Defense were used to divide the US economy into more than 250 different industrial sectors.

After the economy was divided into these industrial sectors, it was then possible to decide which uses the stockpile would supply in a conflict. The new stockpile goals supply those material needs not satisfied through domestic or reliable foreign production for all wartime, defense-related industrial production, direct DOD procurement, and essential civilian goods production.

Thus, the new National Defense Stockpile goals represent a materials stockpile that will supply all our nation's essential material needs in a war but not our non-essential demands (such as the production of toys, recreational vehicles, or other luxury goods). The focus of the stockpile on essential war-related materials needs, as opposed to non-essential civilian goods production, is the single biggest factor accounting for the change in the stockpile goals. Simply stated, the 1979 goals stockpiled materials for a "guns and butter" world war; the new stockpile goals do not. The stockpile study did show, however, that materials *will* be available to produce more luxury or non-essential goods in wartime but that such production will depend upon whether capital and labor are available.

In order to calculate which materials the war-related industries consume during the scenario, the Department of Commerce, using census-derived data and various modeling techniques, projected materials consumption patterns for these industries.

The international material supply-demand balance was calculated by the Department of Treasury. The US materials demands as well as international demands were compared with available worldwide supply. If any of America's essential material requirements outstripped available wartime supply—after a number of foreign material demands were satisfied—a stockpile goal was created for the particular material being studied.

Last, in order to ensure a margin of safety in stockpile planning, sensitivity analysis was applied to the study's results.

As a result of the foregoing analysis, the President adopted a new stockpile goal of $700 million. The President also decided to keep a portion of the existing stocks ($6 billion out of $10.9 billion in inventory) as a supplemental reserve. Although this new stockpile is, by orders of magnitude, the largest in the world, it is almost $10 billion less than what was previously thought to be needed. The goal decreased because of the study's focus on true defense and essential materials needs during a war.

The US economy and the defense industrial base now consume an enormous variety of new, high-technology materials. The National Defense Stockpile and all the stockpile studies, however, have focused principally on the materials bought for the stockpile in the early 1950s. As part of the President's stockpile decision, he requested a follow-on study of the new high-technology materials. If any new goals are identified for these materials, they will be stockpiled.

Richard Levine, former Deputy Assistant Secretary of the Navy for Technology Transfer and Security Assistance, has served on the staff of the Secretary of the Navy and the National Security Council.

Stockpiling Strategic Materials: A Policy Assessment

Paul K. Krueger

During the 200-year history of our nation the dependence of the country on foreign sources for raw materials has been a concern for only the second half of that period. Although the Congress recognized the potential problem in the late nineteenth century and numerous boards and commissions made post-World War I recommendations for stockpiling, it was not until the eve of World War II that the first halting steps for establishing a stockpile took place. Most acquisitions for the existing stockpile occurred during the decade of the fifties; most disposals of excess materials from the stockpile occurred during the sixties and early seventies, with a mix of acquisitions and disposals occurring since that time. The shifts in the types of activity occurring signal fundamental policy shifts within the executive and congressional branches in government. Throughout the time period from 1945 to the present the views of the individual departments and agencies have remained the same. What has been altered has been the power each has wielded as administrations and congresses have changed. This paper will trace these interactions over time and make an assessment of the contribution of these shifting policies to the public good and national security of the nation.

A material is considered strategic if it is imported in substantial quantities and is considered critical if its use is essential to industry. To be eligible for inclusion in the stockpile a material must be both strategic and critical. The current stockpile consists of 63 basic industrial raw materials in a variety of forms that is valued at $8 billion. This material is stored at over 100 locations around the country. Over the years materials such as hog bristles, sperm whale oil, and kyanite have been dropped because there no longer is an industrial requirement for the materials, while germanium was recently added to the stockpile because of its importance in modern electronics.

Early History

The first serious consideration to stockpile raw materials did not occur until the 1880s. Up until that time the forests of New England provided the lumber necessary for ships of the line; the coal fields, limestone quarries, and iron mines of Middle America provided the cast iron for cannon and shot; and the newly opened West provided copper, lead, and zinc for small arms and ammunition.

In short, America's national defense rested upon a strong base of domestic mining and manufacturing. The invention of steel dramatically changed this rosy picture.

The good news was that the brittleness of cast iron was replaced by strong steel that could be rolled into large plates. The US Navy took advantage of these properties and built a fleet of coal-burning, steel-hulled ships driven by steam. This completed the transition, started by the *Monitor* and *Merrimac*, from wind-powered, wooden-hulled ships. Unfortunately the United States was now dependent on foreign sources for one key input to the steelmaking process—manganese. Without manganese it was impossible to make steel and the iron ore bodies of America did not contain substantial quantities of manganese. A committee of the Senate held hearings to determine the impact of this and other vulnerabilities brought on by the new technology. One of the recommendations of the committee was to establish a stockpile of manganese ore. Nothing was done.

America entered World War I totally unprepared for industrial mobilization. It took no action during the early war years to expand munition plants, to design and manufacture the new weapons of war (the tank and the airplane) or to plan with industry on how to mobilize rapidly in the event of war. Most of the Allied Expeditionary Force was equipped by our European allies. Having said this, one must also acknowledge the contributions of Bernard Baruch as Chairman of the War Industries Board and of Charles K. Leith as Minerals Adviser to the War Shipping Board. These two men had a profound effect on postwar industrial mobilization policy and stockpile policy of the United States that grew out of their World War I experiences.

Baruch advocated increased reliance on cheaper imported raw materials to fuel a defense industry expansion; Leith felt it necessary to conserve scarce oceanic shipping capability and to encourage less efficient domestic mining operations to expand. During the war Baruch's views seemed to prevail but with little real impact because the war was over before America's industrial might could be brought to bear on the European Front. Leith continued to have an impact, however, because he went on to serve as an adviser to the American delegation to the Paris Peace Conference and then as an advisor to a number of government-sponsored committees and commissions during the twenties and thirties.

As a result of his experience Baruch felt the government should aggressively stockpile a number of commodities that were necessary to create a modern war-fighting machine. The list now was longer than just a single material, and in addition to manganese, included antimony, chromium, and tungsten. Leith, on the other hand, looked for an international solution. He went to Paris believing that access to mineral resources was the key to continuing world peace. He advocated that an economic council be set up under the League of Nations that would ensure an equitable distribution of raw materials to member nations. He returned from Paris disappointed and disillusioned.

By 1921 Leith had switched from his international solutions to domestic-oriented solutions. He advocated continued peacetime reliance on imports for certain materials in order to conserve scarce, low-grade domestic deposits. Stockpiling would be necessary to provide a buffer in an emergency in order to allow these domestic sources to be developed. Unfortunately, the United States adopted the contrary policy of raising tariffs so that these high-cost mines would be protected. With the single exception of tungsten, the tariffs did not maintain a viable mining industry for these critical materials.

Later in the decade, Leith formed a "Mineral Inquiry" under the joint sponsorship of British and American professional societies. This group recommended that the United States accept raw materials as an offset against war debts. These materials, including manganese, chromium, cobalt, mercury, mica, tin, nickel, rubber, tungsten, radium, and coconut shells, would form a stockpile to serve as a form of national insurance. These concepts were surfaced again by Leith in 1934 through his membership on the Planning Committee for Minerals Policy. By this time the military services were listening. Stockpiling of raw materials financed by repayment of war debts or by the barter of surplus agricultural products was supported. In 1935 the State Department endorsed this idea, and in 1936 legislation for stockpiling was introduced in Congress.

Despite this growing support, President Roosevelt stymied all attempts to start stockpiling because of the political climate within which he had to operate. Only after the Munich Conference of 1938 did he withdraw his opposition. Although the Congress authorized $100 million to be spent in 1939 and 1940, only $10 million was appropriated. Finally, in June 1940 Roosevelt asked for and received an additional $60 million in appropriations.

Because of this foot dragging the nation entered the war seriously short of critical raw materials. The Japanese attacks in Southeast Asia cut the United States off from any supplies of tin, rubber and cordage fiber. Throughout the war incredible efforts had to be made to find new sources for these materials. The United States also lost substantial portions of imports of other key materials including chromium, kapok, manganese, tungsten, and antimony. Other materials, such as copper, lead, and zinc, fell in short supply as industrial demand quickly outstripped domestic supply. In the Caribbean the German U-boat fleets specifically targeted the bauxite barges coming from the South Atlantic to Gulf coast ports. During 1942 90 percent of the barge fleet was sunk in enemy attacks.

Postwar Policies

At the war's end everyone knew the value of raw materials and their contribution to industrial mobilization and the national defense. The case for a sensible stockpiling policy no longer had to be made. In 1946, the joint views of industry, the Congress, and the administration came together in the Strategic and Critical Materials Stock Piling Act. The purpose of the act was "to provide for the acquisition and retention of stocks of certain strategic and critical materials and to encourage the conservation and development of sources of such materials within the United States and thereby to decrease and preclude, when possible, a dangerous and costly dependency by the United States upon foreign sources for supplies of such materials in times of national emergency." This congressionally mandated policy remains in effect today.

The 1946 act also took into account the interests of domestic industry, particularly the mining industry. This industry had expanded during the war to meet the increased demand for raw materials. A significant amount of material remained in the inventories of the Metals Reserve Company, a wartime, government-owned and operated company which controlled the flow of materials during the war. Because of the well-articulated concerns of the mining industry, the materials would be transferred to the new stockpile as long as the Department of Commerce certified that there was no shortage of a particular material in the marketplace. Material valued at over $137 million was transferred to the stockpile from these inventories.

The mining industry also looked to the stockpile as a potential source of business for material being added to the stockpile, and as a potential competitor when material was sold from the stockpile. The manufacturing sector viewed the stockpile conversely, as a competitor for material when buying and as an alternate source when selling. To walk the narrow line between these competing interests Congress provided that stockpile transactions "avoid undue disruption of the usual markets of consumers, producers and processors."

The Army-Navy Munitions Board had the responsibility for determining which materials should be included in the stockpile and how much of each material should be acquired. Most of their initial work borrowed heavily from the recently defunct War Production Board. In fact the War Production Board's list of materials was adopted in its entirety. To determine the appropriate quantity to be stockpiled, separate estimates of anticipated wartime supply and requirements were made. The underlying assumption was that the stockpile should be adequate to support a global conventional war lasting five years—in other words, to refight World War II. Beyond these basic assumptions considerable disagreement existed as to the levels of requirements and supplies. The Army-Navy Munitions Board assumed the military could control all vital sea lanes and estimated supplies would be available from both the Eastern and Western Hemispheres. They also made no provision for civilian demand when calculating requirements. The Department of Interior disagreed with both supply and requirement calculations. In particular, they only counted on Western Hemisphere supplies and fully supported civilian consumption levels. In the end an advisory committee was appointed and the early stockpile objectives were based on requirements levels derived from the per capita consumption of materials that actually occurred in 1943 and supplies principally from the Western Hemisphere. (The overly optimistic assumptions made by the military were consistent with their record during the war. By 1943 the War Production Board refused to base their planning on the military estimates of future requirements for raw materials and industrial output.) In 1947 the role of the Munitions Board was replaced by the National Security Resources Board created by the National Security Act.

Appropriations to purchase materials for the new stockpile were rapidly forthcoming; just 16 days after the Stock Pile Act was signed, $100 million was appropriated for purchases. This speed is hard to believe in today's environment of continuing resolutions. This amount, however, was not sufficient for the program to achieve its goal of reaching the stockpile objectives in five years. Until the outbreak of the Korean war and the issuance of NSC-68 the stockpile program remained a matter of some controversy.

Glory Years

NSC-68 was an assessment of the threat capabilities of the Soviet Union and included a three-year program for military and industrial mobilization. The purpose of this program was to raise the level of deterrence in the face of the perceived threat. Building up the stockpile was a major part of that program.

Material was added to the stockpile in every way one might imagine. For fiscal years 1947 and 1948 the total obligation to purchase stockpile materials was only $340 million; for fiscal years 1949 and 1950 the total was $1.1 billion and by fiscal year 1951 over $1.5 billion was obligated in a single year. In addition to direct purchases, material was acquired by using European counterpart currency as authorized by the Marshall Plan; government-owned industrial facilities were sold to the private sector with payment made in materials in lieu of money; government-owned excess agricultural products were bartered for raw materials, and guaranteed purchases were made using the authorities of the Defense Production Act of 1950. One well-meaning Customs Court judge even awarded the stockpile some gem quality diamonds that had been confiscated from someone smuggling them into the country. By 1960 most of the acquisitions for the stockpile had been completed.

Subsequent Activities

In the intervening years stockpile objectives have gone up and down. Stockpile policy has been used as an instrument to assist the domestic mining industry (the lead and zinc programs of the Eisenhower administration) and has been used to keep the industry in check (the price "jawboning" of the Johnson administration). Both the Johnson and Nixon administrations pursued a program of vigorous sales from the stockpile to generate revenues that would offset budget deficits.

In the following figure showing the history of stockpile objectives and inventories since the start of the program, the bars show the aggregate value of all stockpile material at any point in time and reflect changing policy regarding the stockpile. The line shows the aggregate level of actual inventories at any one time and is a measure of how effectively the policy has been carried out. It is clear that implementation of the policy lags the creation of the policy by many years.

There are two focal points for policy in the executive branch. The first is the Office of Management and Budget and the Council of Economic Advisors. When they are powerful, pressures have been generated to reduce stockpile objectives, reduce purchases, and accelerate sales. The recent experience when David Stockman was Director of OMB is a good illustrative example. The National Security Council and the Federal Emergency Management Agency (the successor agency to the National Security Resources Board) form the second focus. Acting in concert they accelerated the program in the early 1950s and reversed the decision to sell off stockpile materials in the mid-1970s. Five other agencies have an interest in stockpile policy, the Departments of State, Commerce, Defense, Interior, and the Treasury. Commerce, Defense and Interior have usually but not always favored larger stockpiles than State and Treasury. In any case stockpile policy is not the major issue of any of these agencies and their roles were chiefly ones of support.

Congress and the private sector have acted in an effective oversight and monitoring role. Senator Malone in the early 1950s held extensive hearings that led to, among other things, the creation

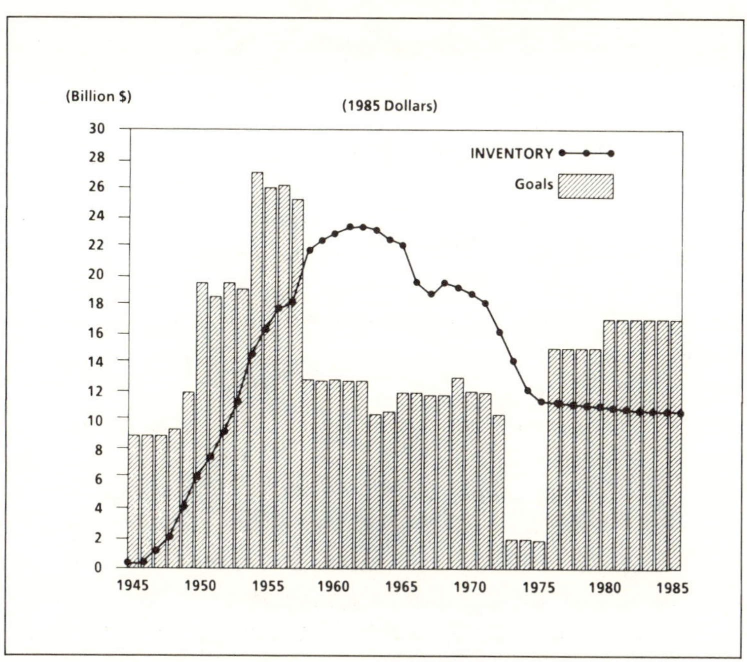

Stockpile Goals Versus Total Inventories.

of the domestic titanium industry through financial incentives for manufacturing and engineering in addition to stockpile purchases. Senator Symington, the former head of the National Security Resources Board, held hearings in the early 1960s that led to an era of much more open stockpile deliberations. Today Congressman Bennett provides an active oversight role and has been a champion of the stockpile since 1975. Supplementing these efforts have been a number of advisory committees staffed by the private sector which have made major contributions to the policy debate. The best known are probably the Paley Commission and the Mott Committee. The formal names of these two groups have been long forgotten, but the two chairmen left their mark.

An Assessment

Over all the stockpile has met its primary objective over the years. During both the Korean and Vietnam wars materials were released from the stockpile that made a direct contribution to the war effort. In addition vital defense efforts were served in 1956 and 1979 when the President made specific releases of material to the Atomic Energy Commission and the Department of Defense, respectively. At other times the material in the stockpile has been loaned on an interim basis to serve national security purposes.

Beyond the direct objectives of the stockpile, the record is less clear. At times the stockpile was a welcome recipient of excess material produced by an expanding mining industry whose civilian customer base had not expanded as rapidly. At other times it has been a difficult competitor in a shrinking market. At times the mere existence of the stockpile has had a beneficial effect on stabilizing prices. Sometimes the mere existence of the stockpile has otherwise deterred new investment in commercial mining ventures.

On balance I believe the existence of a modern, up-to-date stockpile is an important part of the ability to mobilize industry in a national emergency. A rational stockpiling policy, effectively implemented, can be an important part in the equation of national deterrence. However, the key to achieving these benefits is policy stability; without stability the cost of the benefits is very high.

Mr. Paul K. Krueger is the Assistant Associate Director for Mobilization Preparedness Programs at the Federal Emergency Management Agency (FEMA). In this position he has directed, since 1980, the Agency's mobilization preparedness activities for economic, industrial, and natural resource programs, including the stockpiling of strategic and critical materials.

US Strategic Materials Dependence on South Africa

Robert Dale Wilson

Note. Robert Dale Wilson, formerly Director of the Office of Strategic Resources at the Department of Commerce, headed the White House Council on Critical Materials. The following Department of Commerce Paper, prepared under the guidance and direction of Mr. Wilson, examines in detail South Africa's importance to the United States as a supplier of strategic minerals. The paper also stresses the importance of industrial readiness as the linchpin of strategic deterrence in a conventional war environment.

We will never fight World War II again. But that war does have some lessons for us: had we been able to exploit General Patton's initial breakout in France in the fall of 1944, the map of Europe would be different today.

The point is that dramatic strategic outcomes depend on logistical success. In turn, logistical success depends on industrial readiness. In the historical context, there is less and less time to make up for a lack of readiness. Today, industry must be nearly as ready as the infantry.

Industrial readiness is at the bottom end of the long supply lines that support our military. At the foundation of industry lie the minerals and materials from which all things are manufactured. This analysis is about our dependence on South African minerals, those minerals on which we have relied for so many years. It does no good to concern ourselves about these minerals unless we understand clearly our need for them. We need a strategy to ensure their availability to us, and we have to think about the national security dimension of each action we take that has an effect on the minerals and materials on which we depend.

Our dependency on all foreign sources for critically needed minerals, exposed as many of them are to strategic disruption, has grown to 99 percent for manganese, 95 percent for cobalt, 91 percent for platinum and 82 percent for chromium. The Republic of South Africa is our major source of these minerals as it is for vanadium. Although cobalt is not produced in South Africa, it is shipped through South Africa and therefore dependent on their stability. Within this overall foreign dependency, South African shipments of cobalt to the United States represent 61 percent of our consumption; shipments of chromium represent 55 percent; platinum, 49 percent; vanadium, 44 percent; and manganese, 39 percent. These high-dependency rates are cause for alarm if we believe that it will become more difficult in the future to work with the government of South Africa than it has in the past.

Mineral Wealth of the Republic of South Africa

South Africa is blessed with an extraordinary mineral wealth, providing 84 percent of the Western world's manganese metal, 82 percent of the gold, 81 percent of the vanadium, 55 percent of the ferrochromium, 51 percent of the alumina silicates, 42 percent of the platinum group metals, 41 percent of the chromium ore, 40 percent of the manganese ore, 33 percent of the ferromanganese and 31 percent of the diamonds as well as lesser percentages of 55 other commodities. It employs 700,000 people in 656 mines and 434 quarries in the production of these 65 minerals. The importance of these mineral resources was recognized early in the history of the developing nation. In 1910, at the formation of the Union of South Africa, the mining industry "already accounted for some 28 percent of the gross domestic product (GDP)."[1] Through the years, the government has worked jointly with the private sector toward the development of mining, encouraging the mining companies to cooperate together in any matter of mutual interest as a means of reaching higher levels of efficiency. The private sector is thus organized under the auspices of the Chamber of Mines of South Africa, not a government entity. The Minerals Bureau of the Department of Mineral and Energy Affairs is the government arm that works together with the chamber. That high degree of open cooperation among companies and the government has led to full exploitation of the abundant mineral wealth available to the South Africans. Today, a significant portion of the Republic of South Africa's economy is composed of mining and minerals processing, and about 50 percent of its foreign exchange is earned from exports of mineral commodities.

US Dependency

The materials situation in our own country, on the other hand, is moving toward increasing dependency on foreign sources for critical and strategic minerals; some are associated with the Republic of South Africa.

Chromium. The United States imports 82 percent of its chromium requirements; the remainder is supplied from government and industry stocks and stainless steel recycling. No domestic producers of primary chromium exist at this time. The European Community and Japan are more than 90 percent dependent on imports of chromium for their domestic needs.

The Republic of South Africa accounts for about 31 percent of world mine production of chromium; the Soviet Union accounts for 35 percent; and Albania furnishes 9 percent, for a total of 75 percent from these three countries. Of US imports of chromium, 72 percent are from the Republic of South Africa. This situation is expected to continue, as the United States has no known reserves of shipping grade chromium, and the Republic of South Africa has 84 percent of the world's chromium reserve base. Demand for chromium in the United States has increased from the reduced levels of 1983 and is expected to continue to increase through 1990.

World chromite production is only about 74 percent of available capacity, therefore no shortages are expected to develop under current conditions. However, in the event of a cutoff of South African (including Zimbabwe) chromium supplies, excess market economy production capacity, excluding those two countries, would amount to only 36 percent of South African (including Zimbabwe) production in 1984, very likely resulting in a supply shortfall within the market economy countries and greatly increased chromium prices. This assumes that the rest of the market economies would be perfectly efficient in their ability to produce at their full 100 percent capacity and distribute in precise amounts to those users who were denied African or Zimbabwean chromium. Any market inefficiencies would exacerbate the loss, with attendant price increases.

In the short term, the foregoing losses could be offset by a combination of actions: substitution among products where a loss in performance would be acceptable, an increase in recycling, and the use of current private producer and consumer chromium stocks that are believed to amount to approximately 144,000 short tons of contained chromium. Based on these circumstances, chromium demand could be reduced to an estimated 306,000 short tons of contained chromium annually, 103,000 short tons of which could be supplied through recycling and about 100,000 short tons of which could be supplied by current alternative market economy import sources, leaving a shortfall of 103,000 short tons on an annual basis. Current private producer and consumer stocks would be exhausted within the first 14 months of the cutoff, and severe shortages could then be expected to occur.

Should a national emergency occur, chromium in the National Defense Stockpile would be made available to domestic producers. Under President Reagan's proposal, the stockpile would contain 793,423 short tons of chromium metal equivalent. However, the chromium metal currently in the stockpile is not of sufficient quality for the production of vacuum-melted superalloys for aircraft jet engines or other alloys that have stringent purity requirements, such as titanium alloys and molybdenum stainless steel, both of which have important defense applications.

Manganese. The United States imports 99 percent of its manganese requirements, possessing no manganese deposits of sufficient grade to warrant production. The US imports consist of both manganese and ferromanganese. The European Community and Japan import 99 percent and 95 percent, respectively, of the manganese required to supply their domestic needs.

The Republic of South Africa will continue to be an important producer of manganese, possessing 71 percent of the world's manganese reserve base, as will the Soviet Union, with 21 percent of the reserve base. By comparison, Australia possesses only 5 percent of the total world reserve base of manganese. Significant quantities of manganese occur on the deep ocean floors, but they are not economically recoverable at this time.

Manganese consumption in the United States is expected to increase at a rate of about 3 percent annually through 1990, following the trend in raw steel production. World manganese production capacity utilization amounted to only 70 percent in 1984. The current excess of 1.5 million short tons of market economy manganese production capacity (excluding South Africa) could offset a cutoff of manganese from the Republic of South Africa, with production of only 1.3 million short tons in 1984. Private produce and consumer stocks would further mitigate a cutoff of South African manganese supplies to the United States. Some price increases could be expected as consumers sought new suppliers.

In case difficulties are encountered in expanding production to earlier levels, or should the rate of consumption of steel rise faster than expected, shortages could develop. If so, manganese production could be increased in Australia within one to two years, and in Brazil, India, and Mexico within a slightly longer time period. However, such increases are not being planned at this time and should not be expected to occur without government assistance. Development of manganese resources in the United States would take at least three years and would be extremely costly.

The best opportunity for the United States to reduce its vulnerability to manganese imports is through technological innovation. The development of better process controls in steelmaking could reduce the requirements for manganese by about 8 percent. Manganese used in steel production could also be reduced by external desulfurization. However, both of these innovations will not become widespread until the 1990s.

In the event of a national emergency, manganese contained in the National Defense Stockpile would be available to domestic producers. President Reagan's stockpile goal of 869,667 short tons of contained manganese would supply US producers for 14 months under current consumption patterns.

Platinum-group metals. The platinum-group metals (PGM) consist of platinum, palladium, iridium, osmium, rhodium, and ruthenium. The United States imports approximately 90 percent of its PGM requirements. Small amounts of PGMs are produced domestically by copper refineries as by-products and by recyclers. The European Community imports 100 percent of its PGM requirements, and Japan imports 95 percent.

The Republic of South Africa accounts for about 41 percent of world mine production of PGMs while the Soviet Union accounts for about 52 percent. Of US imports of platinum-group metals, 60 percent was supplied by the Republic of South Africa.

The Republic of South Africa possesses 81 percent of the world's PGM reserve base, while the Soviet Union possesses 17 percent. The United States has a small reserve base (1 percent of the world total) located in the Stillwater Complex in Montana. Production from this deposit is currently underway. It is expected to produce about 2 percent initially, building up eventually to 6 percent of domestic platinum and palladium consumption.

World PGM production amounts to 79 percent of available capacity. Total world capacity is currently 8.9 million troy ounces annually, but market capacity excluding South Africa is only about 900,000 troy ounces. Apparent annual consumption in the United States is about 2.6 million troy ounces of platinum-group metals and is expected to increase about 2 percent per year through 1990. Therefore, a cutoff from South African PGM supplies could result in severe market disruption and would provide the Soviet Union with a great deal of market leverage with its 4 million troy ounces of production capacity.

Given such a disruption, US private producer and consumer stocks would supply the market for about 6 months. Domestic primary PGM supply could be increased by accelerating Stillwater production; however, any increase over 2 percent of consumption would be costly. Substitution in most uses would reduce performance and would most likely only occur with an extreme increase in price. Conservation by increased recycling of catalytic converters in scrapped automobiles could reduce import dependence, but some barriers exist that could delay any significant increase for several years at least. For example, some recycling processes are proprietary and the number of collection centers for catalytic converters is limited. Many automobile scrap dealers fail to remove the converter prior to scrapping, reducing the supply considerably. Under current market conditions, recycling of catalytic converters is estimated to provide an additional 300,000 troy ounces of platinum, palladium, and rhodium by 1995.

At the announcement of a national emergency, consumption could be reduced substantially by removing emissions standards. Discontinuing the use of catalytic converters in automobiles could reduce consumption by 35 percent to about 1.6 million troy ounces per year. Further, recycling efforts could be stepped up, but the paucity of recycling facilities could result in severe bottlenecks. With current primary and secondary production, development of the Stillwater Complex, emissions standards, a shortfall of supply of about 1 million troy ounces of PGM would result based on 1984 consumption levels after the first year of a cutoff of imports. President Reagan's National Defense Stockpile goal for platinum-group metals was zero.

Vanadium. The United States imports 54 percent of its vanadium requirements. Vanadium oxide is produced domestically from Colorado Plateau uranium-vanadium ores and Idaho ferrophosphorus slags. Oxides of vanadium are produced domestically from petroleum residues, utility ash, spent catalysts, and imported iron slags. The European Community is 100 percent import dependent for vanadium, and Japan imports 70 percent of its vanadium requirements.

The Republic of South Africa accounts for 40 percent of world mine production of vanadium, the Soviet Union for 31 percent, the People's Republic of China for 15 percent, Finland for 10 percent, and the United States for 5 percent. Recently, 38 percent of US imports of vanadium were from the Republic of South Africa, 25 percent were from the European Community (mainly produced from South African raw materials); 16 percent were from Canada (also produced mainly from South African raw materials), and 6 percent were from Finland. South Africa will likely remain as an important producer of vanadium because of its large reserve base totaling 17.2 million pounds of contained vanadium, or 47 percent of the world's total. The Soviet Union also has a large reserve base, with 25 percent of the total. The United States possesses 13 percent of the world's vanadium reserve base.

Apparent consumption of vanadium in the United States amounts to 11.4 million pounds and is expected to increase at a rate of 3 percent per year through 1990. Consumption increases are expected as a result of greater demand for high-strength low-alloy steels. Vanadium production from domestic ores, concentrates, slags, and petroleum residues in the United States was about 8.6 million pounds. Five of the 12 US vanadium mining facilities are on standby and mill capacity utilization is just over 50 percent. Vanadium recovered from mill products is greater than mine production because mills processed stockpiled materials. World capacity utilization is about 2 percent.

We receive about 28 million pounds of contained vanadium from South Africa each year. Should we experience a cutoff of those supplies, market economy excess capacity of 23 million pounds of contained vanadium could not satisfy demand under current conditions. Although the United States has sufficient capacity currently to supply its own market at higher than current prices, the European countries would likely suffer severe shortages. One factor which would mitigate the effects of a cutoff in the United States is the existence of large private producer and consumer stocks, amounting to 7.5 million pounds of contained vanadium. In the event of a national emergency, the 1.4 million pounds of vanadium metal equivalent in the National Defense Stockpile under President Reagan's proposal would become available for domestic use. However, about 75 percent of this material would require upgrading prior to use.

Cobalt. The United States imports 95 percent of its supply of cobalt. Minor amounts of cobalt are produced domestically from scrap. Primary cobalt has not been produced in the United States since 1971. The European Community and Japan are both 100 percent dependent on imports to supply their requirements for cobalt.

Although the Republic of South Africa produces no cobalt domestically, cobalt produced in Zambia and Zaire is transported through South Africa for export to the United States and other countries. Zaire produces about 55 percent of total world mine production of cobalt, Zambia 15 percent, the Soviet Union 8 percent, Canada 6 percent, Cuba 5 percent, and Australia 4 percent. Zaire supplies about 37 percent of US cobalt imports, Zambia 12 percent, Canada 10 percent, and Japan 7 percent. A total of about 70 percent of the cobalt consumed in the United States is transported through the Republic of South Africa.

Today, the United States produces no primary cobalt. We do have a reserve base of 950,000 short tons of contained cobalt. These reserves are subeconomic at the present time. Zaire and Zambia will both very likely continue to be important sources of cobalt, with a combined reserve base of 2.9 million short tons of contained cobalt, or 32 percent of the world cobalt reserve base. Cuba's cobalt reserve base amounts to 22 percent of the world total, and New Caledonia and the United States each account for 10 percent of the world total.

The US demand for cobalt is expected to increase at an average annual rate of 3 percent through 1990. Under current economic and political conditions, no shortages are expected to develop. World cobalt production amounts to only 80 percent of total capacity. However, should the political situation in the Republic of South Africa cause a disruption in the transportation system, shipments of cobalt from Zaire and Zambia could be severely curtailed. The only easily accessible, alternative transport route from these countries is by air. This has been done in the past, but the result was substantially increased cobalt prices. The shortfall of production from Zaire and Zambia could not be met by excess market economy capacity, which amounts to less than one-third of production from these two African countries.

In the event that cobalt prices did increase significantly, some of the US cobalt deposits could be brought into production, particularly the Blackbird Mine in Idaho and the Madison Mine in Missouri. After a development period of two years, these two mines combined could produce up to 3,000 short tons of cobalt per year or possibly one-third of US consumption.

In the short term, producer stocks could supply consumers for about seven months. Substitution potential for cobalt in most uses is not good, and therefore price increases would most likely be the result of transportation system disruptions. In the event of a national emergency, cobalt in the National Defense Stockpile under President Reagan's proposal would supply domestic consumers for a period of about 1.5 years at current consumption levels.

Interagency Field Study of South African Minerals

Based on concern for all of the foregoing, an Interagency Field Team from the Departments of Commerce and Defense conducted a field study in the Republic of South Africa to gain first-hand knowledge of the then current status and outlook of the South African minerals industry. This trip provided the interagency community responsible for the formulation of national minerals and materials options with more information on the reality of the situation in South Africa as it relates to our materials dependency. The team produced some recommendations. First, that the interagency community evolve a comprehensive analytical model for relating those factors affecting vulnerability with those affecting commercial and security interests. Second, that the role of the Republic of South Africa in supplying minerals to the United States and its European and Japanese allies must be assessed on a continuing basis, not just from reported data but from firsthand observation as well. Third, generalizing from the South African experience, on the supply side there is a need to develop a methodological approach for assessing vulnerability and dependency patterns affecting the industrial base (and readiness) of the United States under a number of planning scenarios covering the spectrum of likely contingencies over the next 5 to 10 years. As a foundation, this effort would trace the flow of minerals and materials from the original point of production, through their various transformations where value is added to the ultimate US and allied users. This basic flow process analysis would provide the core information for policy formulation in the interagency community in the areas of emergency preparedness, industrial mobilization, stockpile requirements estimation, and other aspects of national resource allocation planning for contingencies.

On the demand side of the minerals flow, the team recommended that it is essential to gain the current perspective and long-range outlook of the private sector and, in particular, to determine the potential private sector impacts from the supply side contingencies.

The team also recommended, first, assessing the potential for expanding supplies from other new or existing sources; second, initiating parallel analyses with other suppliers; third, coordinating combined action among allies in the international community; and fourth, assigning responsibility for leading this interagency effort to the Office of Strategic Resources in the United States Department of Commerce that stimulated the initiative originally.

The Anti-Apartheid Act of 1986

Section 502 of the Anti-Apartheid Act of 1986 (PL 99-440) required that (1) a baseline report be prepared showing imports of strategic and critical materials from Member and Observer countries of the Council for Mutual Economic Assistance (CMEA) during the years 1981 through 1985 and (2) that current reports be prepared each month showing comparable import data. Since the act was passed, current reports have been prepared and furnished to the Congress. The purpose of these reports is to permit comparison of the monthly reports with average monthly data during the baseline period. The act provides that, under certain circumstances, the President can remove sanctions imposed elsewhere in the act if current monthly imports exceed baseline average monthly data for comparable materials.

Comparing current reports between October 1986 and July 1987, a number of imports of critical materials from CMEA have increased over their average imports during the baseline period—specifically from the USSR, Yugoslavia, Poland, and Czechoslovakia.

The most significant increases in imports by the United States from these countries are for antimony, chromite (refractory grade), diamonds, ferrosilicon, platinum bars and rhodium from the USSR and for ferrosiliconmanganese, silver, and zinc from Yugoslavia.

The increases from the USSR have been as follows:

—*Antimony*. The monthly average imports during the base period were only 1,281 pounds. Shipments since October 1986 have averaged as much as 98 times this amount. This increase coincided with a rapid expansion of antimony production in the Soviet Central Asian republics of Kirgiziya and Tadzhikistan.

—Chromite (refractory grade). The Soviets have expanded the Donskoye chromite complex in Kazakhstan where the new Molodezhnaya Mine with a capacity of 2 million tons per year of ore is being developed. Coincidentally with this development, the United States passed the Anti-Apartheid act. Average imports during the months since the Act have been 13 times those received during the base period.

—*Diamonds (industrial)*. The 1981-85 baseline average was only 2 carats; the January 1987 shipment was 1,200 carats.

—*Ferrosilicon*. Increasing imports beginning last December have resulted in current averages that have been four times that generated during the base period.

—*Platinum-group metals*. PGM imports are down 19.1 percent overall as a result of decreases in shipments of ruthenium and palladium. On the other hand, platinum bars and plates were up 5 times, rhodium up 3.5 times and platinum sponge imports up 86.7 percent.

Shipments from Yugoslavia are as follows:

—*Ferrosiliconmanganese*. The average shipments during the period of October 1986 through March 1987 were *up 1.5 times* the monthly average for the base period. Much of this material is transshipped through Yugoslavia.
—*Silver*. During the period of October 1986 through July 1987, imports exceeded the average monthly shipments during the base period by *2.33 times*. This silver is produced as a by-product of lead and zinc smelting. Yugoslavia has just recently expanded this capability.
—*Zinc*. Imports for this period were *up 4.5 times* the base period average.

Increases in shipments from Czechoslovakia were are follows:

—*Platinum-group metals*. New shipments of platinum sponge and palladium were received from Czechoslovakia in October and November 1986 where there had been none during the base period.

Some of these materials have critical uses, as described in *Mineral Facts and Problems*, 1985 Edition, Bulletin 675, US Bureau of Mines:

Antimony. The largest use (65 percent of current domestic consumption) of antimony is as a constituent in flame retardant compounds. The three main flame retardant compounds are antimony trioxide, antimony pentoxide, and sodium antimonate. These compounds are used to improve the flammability resistance of plastics, paints, textiles, and rubber, all of which have military applications.

The infrared-reflecting characteristics of certain antimony pigments have led to their use in camouflage paints. Antimony is also used in various ammunition components. Alloyed with lead, it hardens small-arms bullets and shot. Some tracer bullets have a recess in the base containing a light-emitting antimony sulfide mixture that permits visual tracking of the projectile. Burning antimony sulfide creates a dense white smoke that is used in visual fire control, in sea markers, and in visual signaling. Antimony sulfide is also used in percussion-type ammunition primers.

High purity antimony metal (99.999 percent) is used as a dopant in n-type semiconductor material and in the manufacture of intermetallic semiconductor materials such as indium antimonade, aluminum antimonide, and gallium antimonide. Sodium antimonate is used as a fining agent and a decolorizer in specialty glasses for cathode ray tubes used as visual monitor screens for computers, radars, and sonars.

Antimony metal is also used as an alloying element in lead-acid storage batteries, power transmission equipment, type metal, solder, and chemical pumps and piping.

Refractory chromite. The major application of chromite refractories is in iron and steel processing, nonferrous alloy refining, glassmaking, and cement processing. Chromite sand is used to make molds for ferrous castings.

Refractory chromite, while containing less chromium (40 percent or less) than chemical grade (41 to 46 percent) or metallurgical grade (46 percent or more), can be used in place of the other two grades. (It is believed that the shipment from the USSR was intended to be used to produce ferrochromium. Ferrochromium can be used directly to produce various types of alloyed and specialty steels or it can be processed into high purity chromium metal for use in producing all types of superalloys.)

Ferrosilicon and ferrosiliconmanganese. Ferrosilicon and ferrosiliconmanganese are used in the production of various alloyed and specialty steels. Such steels are used as hull plates for naval vessels and in the bodies of military vehicles and tanks.

Platinum-group metals. The largest uses of platinum group metals (67 percent) are as catalysts in automotive catalytic converters, in the organic, inorganic and petrochemical industries, and in petroleum refining. Three-way automotive catalysts usually contain 0.05 troy ounces of platinum, 0.02 ounces of palladium and 0.005 ounces of rhodium. Chemical catalysts are generally platinum, palladium or a combination of platinum, palladium, and rhodium. Petroleum catalysts are usually palladium, for hydrocracking, bimetallic platinum and rhenium or platinum and iridium for reforming, and platinum for isomerization.

The PGMs are also used in the electrical and electronic industries. Thin films consisting of layers of platinum or palladium are used to provide conductor adhesion to electronic circuits. In thick-film circuits, platinum and palladium alloys are applied as pastes and used as conductors. The metals platinum, rhodium, and iridium are used in thermocouples and furnace heater windings; palladium and silver alloys are used in capacitors.

Other uses of PGMs include ceramic and glass production equipment, in various dental and medical appliances (e.g., electrodes in cardiac pacemakers), and in jewelry (much more popular in Japan and Western Europe than in the US).

Silver. Silver is used in electrical and electronic products because of its high electrical and thermal conductivity and its resistance to corrosion. Silver has the highest thermal and electrical conductivity of any known substance. Pure silver is generally used in low- and medium-current switching devices. In other applications where the device requires higher strength, more wear resistance, better resistance to arcing, or lower costs, silver is usually alloyed with another metal such as copper or palladium to produce the desired characteristics.

Silver batteries are used in defense and space applications where battery weight, size, and reliability are major concerns. Silver-zinc button cells are used in calculators and hearing aids. The largest industrial use of silver is in photographic materials. It is used in the manufacture of film, photographic paper, photocopying paper, x-ray film, photo-offset printing plates, and in some other light sensitive products.

Silver is also used in mirrors, catalysts, medicinals, dental amalgams, bearings, refrigeration/air conditioning brazing alloys, and in various decorative and commemorative applications.

Zinc. Zinc metal is used in numerous corrosion protection systems. Zinc plates are used on ship hulls to prevent cathodic corrosion of steel hull plates. The coating of steel products with zinc, i.e., galvanizing, provides two types of corrosion protection. First it provides a long-lived barrier by preventing contact between the steel base and its corrosive surrounding, and secondly, when broken or corroded through, it protects by galvanic action.

Zinc is alloyed with copper to produce brass for use in ammunition and for decorative purposes. Zinc compounds, primarily oxide, are used in anti-corrosive paints, as an accelerator and activator in vulcanizing rubber, as a chemical in photocopying, in inks, dyes, oil additives, wood preservatives, fungicides, varnishes, and phosphors for cathode ray tubes (computer monitors, radar and sonar scopes).

Conclusion

Dependency on foreign sources of materials critical to the US industrial base is not a problem per se. If such foreign sources were to be disrupted by natural disasters or social or political activities, dependency could become a vulnerability. To guard against such vulnerabilities, the United States has pursued stockpiling; research and development on conservation (e.g., near net-shape casting), recycling, and substitution; and development of alternative sources. The priority given such actions depends upon the perceived vulnerabilities of disruption. As conventional forces become more important as a deterrent to world conflict, an even higher priority will have to be given to programs designed to reduce our vulnerability to critical and strategic materials imports.

Notes

1. The Minerals Bureau of South Africa, *South Africa's Mineral Industry*, 60 Juta Street, Braamfontein 2001, August 1984, p. 2.

Robert Dale Wilson, appointed Executive Director of the National Critical Materials Council in June 1987, is responsible for direction and coordination of resource activities and policies within the various operating agencies of the federal government.

5

In the Final Analysis

THE SURVEY OF EXPERT OPINIONS IN CHAPTER 4 SHOWED MORE points of agreement than disagreement. Where the analysis diverged most, it was due more to a matter of emphasis rather than disagreement over the basic issues involved. In many cases, differences of opinion were subtle, reflecting primarily the corporate or government agendas of the respondents. As expected, private industry was more concerned with the forces of supply and demand, markets and prices, and materials needs. Government respondents were more concerned about policies and their lack of consistency. The need for a streamlined, modern, and flexible stockpile, consistent with changing needs, was a common thread among all respondents, as was the view that traditional, hard-metal materials are giving way to advanced high-tech materials.

The experts who were surveyed made several major points:

—There is a high degree of dependence by the United States on foreign suppliers of strategic materials—especially on South Africa—as well as on fabricating capacity.

—Views vary as to whether this dependence constitutes a high degree of risk for the United States.

—Chromium and the platinum group metals present the greatest risks for the United States because South Africa and the USSR control most of the world's resources of these metals.

—The Resource War theory is implausible for economic and political reasons and, thus, does not present a real threat to US national security.

—Private industry is actively pursuing programs in the areas of substitution, conservation, recycling, and new processing technologies in order to conserve scarce supplies of strategic minerals.

—New alloys with substantially lower amounts of strategic metals are being successfully developed, but lead times are long and costs high.

—Advanced high-tech materials, not contained in the National Defense Stockpile, are replacing conventional materials in weapons systems, especially in electronics applications.

—A modern, up-to-date stockpile, not subject to political or economic manipulation, is an effective form of insurance against supply cutoffs or the need to increase industrial production rapidly.

—The US industrial base has eroded sharply due to its lack of competitiveness with overseas producers.

—The Department of Defense has plans and programs underway to assure an ability to mobilize the industrial base rapidly in the event of military emergencies.

—The National Security Council and the Office of Management and Budget stood alone in their belief that President Reagan's recommendation to drastically reduce inventories of stockpiled strategic minerals would, nonetheless, provide an adequate safeguard against potential supply disruptions or the need to surge industrial production.

—Congress has an important oversight role in stockpile matters and intends to exercise this responsibility forcefully to prevent manipulation of the stockpile for political or budgetary reasons.

—Lack of policy stability and foresight, as well as fragmentation of responsibilities among various government agencies, have been primarily responsible for whatever inadequacies exist in the present inventory of stockpiled materials.

Strategic Minerals Policy: A Retrospective Look

The preceding chapters have laid out in some detail the complexity of the strategic minerals problem as seen by the author, as well as by experts both in the private and public sectors. Strategic minerals vulnerability is not a new issue but one over which the US Government has been grappling over for seven decades. Many approaches have been tried in order to come to grips with the issues of dependence and vulnerability. Indeed, a "panoply" of partial solutions have been developed and implemented—some of which have been successful, others less so. However, none has provided an integrated, well-thought-out plan of attack to deal with the problem.

Among the major policy failings have been lack of coherence, consistency, and sustainability. Also contributing to the halting progress of positive initiatives has been a fragmentation of authority, as well as the vested interests of various departments of the executive branch and the Congress. Vacillation by past administrations often has undermined much of the expert analysis done at lower levels. In addition, the ideas, expertise, and advice of the private sector either have been excluded entirely from policy formulation or—when sought—subsequently ignored. Partisan politics has, in more than one instance, skewed expert analysis in order to achieve nonstrategic ends, subjugating the strategic interests of the country to the public good in other arenas. Both Congress and the executive branch must share a portion of the blame. As a result of these negative factors, the United States is little closer to a "solution" to the problem of strategic minerals dependency than it was decades ago.

Perhaps the greatest failing of past policy has been its reactionary nature in the face of unexpected crises. This phenomenon permeated policy during World War I, World War II, the Korean conflict, Vietnam, the 1978-79 cobalt crisis, and more recently, the political crisis in South Africa. Based on this history of reactionism, as well as on a lack of well-defined policies, one could reasonably expect a similarly-reactive approach to the next crisis—be it a disruption in strategic mineral supplies from South Africa, Soviet interference—either directly or through its proxies—in southern Africa, or other unforeseen events. In short, the United States has no apparent pro-active integrated plan to deal with such emergencies.

A New Approach to the Problem

An entirely new approach now needs to be taken. The model described below is based on one developed in July 1986 at a National Defense University workshop dealing with the development of a White Paper laying out US national strategic objectives. The salient elements of the model—adapted here to US strategic minerals policy—are

—Develop a statement of US interests, objectives, and goals vis-a-vis strategic minerals.
—Prioritize these objectives.
—Define in concrete terms a set of prescriptive policies consistent with the achievement of these goals and interests.
—Agree on a specific time horizon for achieving these objectives.
—Simultaneously, work to achieve a national, nonpartisan consensus among cabinet-level officers and Congress—without which the national objectives cannot be effectively carried out.
—Develop specific strategies, both foreign and domestic, designed to achieve the delineated policy objectives.
—Identify the tools or means available to implement the specified strategies, i.e., reconcile means with ends.
—Lay out specific responsibilities within the government for carrying out the national strategy and for applying the implements determined to be most effective within each agency's area of responsibility.
—Develop milestones consistent with the achievement of long-term objectives.
—Require accountability in the form of regular reporting—by responsible officials—concerning what they are doing to implement national policy and how much progress has been made.

—Finally, develop a set of contingency plans capable of dealing with possible crises, especially for those which, while perhaps unlikely, would have a great deal of impact if they did occur.

Under this construct, strategic minerals policy, risks, vulnerabilities, and solutions would be elevated to a national level of importance, rather than left to simmer "on the back burner." By having a well-defined plan, there would be no doubt within the various arms of government regarding US objectives and how the United States planned to meet these objectives. Of most importance, such an approach would create—for the first time—a prescriptive rather than reactionary policy—a much more effective means of dealing with any important problem.

Although implementation of this model would not necessarily be easy, it is, nonetheless, doable. Indeed, difficult problems demand extraordinary effort. But one of America's great assets is its human capital. For example, Lee Iacocca—albeit with government assistance—turned the Chrysler Corporation from a financial basket case to one of America's great success stories. Similarly, within government, the CIA is well-known for its "can do" approach to difficult problems and has been able to sustain that approach over the years by utilizing the best minds available. Elsewhere, both government and private industry are replete with innovative, goal-oriented, self-starters. Indeed, the most successful corporations are headed by men and women with strong leadership qualities who know how to succeed where others have failed. The US Government should be no exception.

A Radical Prescription for Action

The above discussion brings to the forefront the need for major, perhaps even radical, changes in US strategic minerals policy. Because of the current fragmentation of responsibilities for strategic minerals analysis and the vested interests of the many agencies still involved—agencies often working at cross-purposes with one another—an *independent or quasi-governmental office should be established* whose sole purpose is the development and implementation of US strategic minerals policy. Its director would be nominated by the President and approved by the Congress, and it would have ultimate authority over developing and implementing US strategic minerals policy. Admittedly a controversial approach, I would, nonetheless, argue that by consolidating strategic minerals responsibilities within one nonpartisan office, vested interests and disparate agendas would be largely eliminated and the public interest better served.

Staffing such an organization would be relatively easy, considering the literally hundreds of private and public sector experts now working various parts of the problem. Ideally, this new office would draw from a core of analysts who are experts in technology, resources, economics, foreign affairs, military and industrial mobilization, and intelligence. It would report directly to the President's staff—submitting policy recommendations for his approval—while sharing its "action plans" with the Congress. Its director should serve no less than six years—to keep the new agency from becoming politicized by every Presidential election. Also the agency would have power and authority over other government agencies to direct them to carry out the President's strategic mineral policy initiatives. Such a reorganization would at the same time relieve other agencies from policy-making responsibilities in strategic minerals, allowing them to continue to carry out their primary responsibilities in other areas. As a by-product, parochial interests and turf battles would be virtually eliminated.

Putting the Model to Work

The following hypothetical plan briefly illustrates how the model might be put into action.

National strategy objectives. To reduce US import reliance for designated strategic minerals to 50 percent or less by 1995, to foster the development of domestic resources, and to increase foreign strategic mineral supplies from allies and other reliable countries.

Domestic action. To increase US production of designated minerals

—intensify exploration, using satellites as tools of geologic inference; utilize oil company experience in locating promising mineral formations;

—provide industry with investment tax credits to further explore, develop, and produce the designated strategic minerals;

—re-invigorate the defense-industrial base, using economic incentives to increase domestic production efficiency—especially in the minerals processing and fabricating industries;

—increase research and development efforts in order to find adequate substitute materials—using both government-funded programs and tax incentives for private industry; set up an information data base on materials properties;

—provide tax incentives to recycling firms and firms that conserve strategic minerals in producing their final product;

—require private firms to carry working inventories equivalent to one year's annual production, with appropriate government subsidies if necessary; and

—require OMB and Treasury to develop a plan to finance the above tax incentives.

DOD and the National Defense Stockpile. To make the National Defense Stockpile responsive to the nation's strategic needs

—require DOD to prepare and adopt an integrated plan of mobilization based on sustainability during a conventional war of unlimited duration—Figure 12 shows one such model developed by the Mobilization Concepts Development Center of the Institute for National Strategic Studies at the National Defense University;[1]

—require the Defense Department to reconcile its material needs with planned increases in its procurement of planes, ships, tanks, artillery, etc.; require DOD to calculate from the ground up the quantity of materials needed to produce each of these items and to sum these materials requirements;

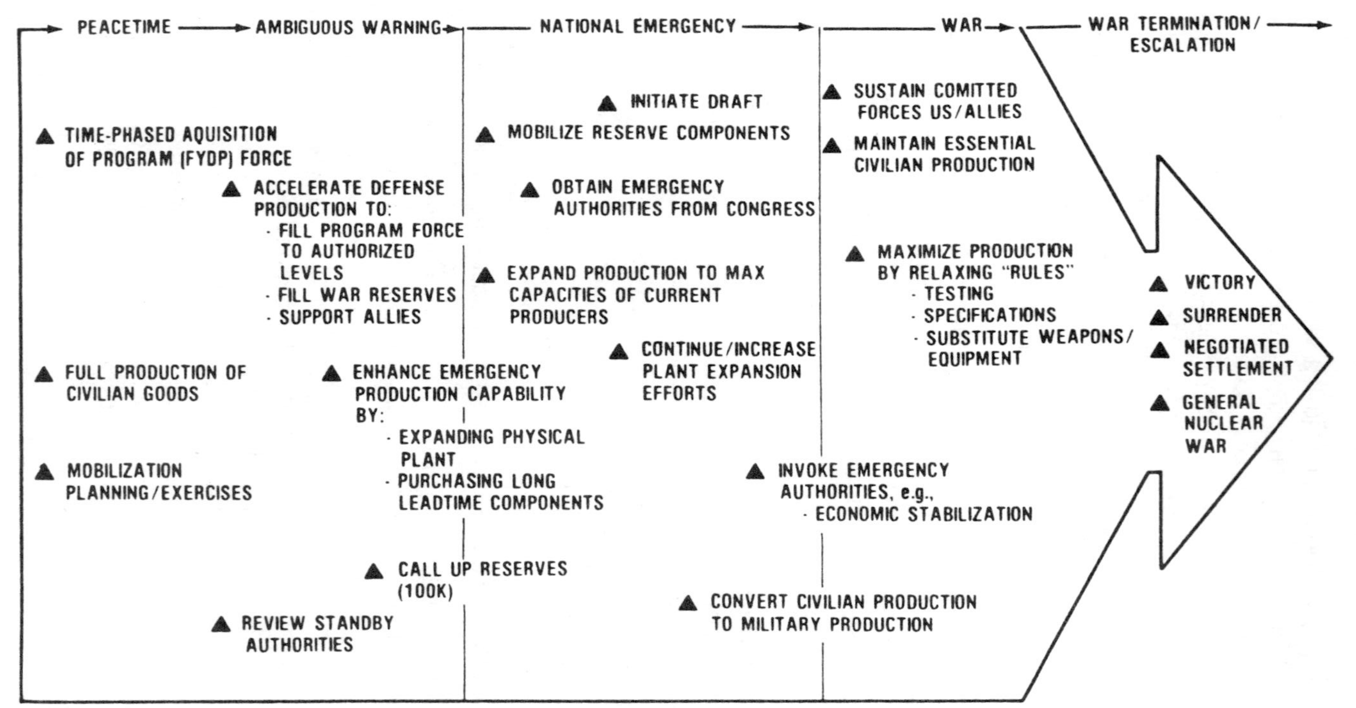

Figure 11. Mobilization Spectrum.

—set up a plan to begin substituting domestically produced material, where possible, for foreign sourced supplies in filling stockpile goals; and

—keep Congress well informed during policy formulation; submit realistic appropriations requests for stockpile purchases and maintain revenues from stockpile dispositions in a fund separate from the General Treasury Fund; use the strengths of key Congressmen—i.e., those that support and understand the strategic interests of the United States—and their abilities to persuade others within Congress that well-defined national security objectives are in the public's best interest.

Foreign policy. To integrate strategic materials objectives with other foreign policy objectives among friends, allies, and the Third World

—provide clear guidance to State, AID, and other internationally oriented agencies as to the nature of US strategic mineral objectives;

—develop joint programs with Canada, Mexico, and Brazil to explore for and produce strategic minerals; provide incentives for private industry to invest in strategic mineral mining operations, especially in Brazil where Japanese investment is now dominant;

—provide development assistance to Third World countries for the sole purpose of increasing the production of strategic minerals; for example, foster the development of cobalt resources in the Philippines and access to Brazil's huge manganese resources—perhaps granting agricultural concessions on a quid pro quo basis—and give US technical assistance to the under-explored countries of central and northern Africa;

—use bilateral negotiations with Pretoria to reconcile clearly US needs for South Africa's strategic minerals with US foreign policy concerning the issues of apartheid and political change within that country;

—establish a jointly developed plan with the PRC—a country of vast untapped mineral wealth—that would make US mineral processing technology available in exchange for supplies of certain critically needed strategic minerals;

—during US-USSR bilateral negotiations, raise the issue of Soviet intentions in southern Africa vis-a-vis strategic minerals, especially as they relate to Soviet-sponsored interference within South Africa and the Frontline States;

—more vigorously support efforts by southern African states to develop the infrastructure necessary to decrease their reliance on South Africa's transportation network in the areas of mining industry equipment, fuel imports, and mineral export shipping capabilities; and

—persuade, coax, or cajole the NATO Alliance to shoulder more of the burden of defense as it applies to the need for strategic minerals to carry out a conventional war. Require the NATO countries to increase their stockpiling efforts to certain minimum levels. Negotiate a quid pro quo agreement on sharing our stockpiled metals in return for access to their metal fabricating and processing capacity.

Crisis management. To identify possible political or military crises that would impinge on our access to supplies of strategic minerals abroad:

—Develop contingency plans in the event any of these crises materialized.

—Rebuild our merchant marine to a level capable of shipping ores and processed minerals to and from Europe in the event of a conventional war.

—Require DOD and the intelligence agencies to identify possible strategic minerals targets—mines, processing plants, transportation chokepoints, etc.—in the event of the outbreak of a conventional war with the USSR.

—Develop a set of options for the use of covert action to (a) reduce enemy capabilities to produce and supply their own war effort with strategic minerals, and (b) to counteract possible interference by them in the supplies of strategic minerals coming out of southern Africa and Asia necessary to support our own war effort.

Conclusions

No one should suffer delusions that all of these are original ideas or that there is only one way to proceed. Indeed, many of the illustrative recommendations are but a synthesis of the ideas of others, some of which are already being implemented, if in a fragmentary way. The major failing has been the lack of a set of national strategy objectives and a unified plan to achieve these objectives. Without a national plan of action, progress has been haphazard, sputtering, and far from what is achievable. Also sorely lacking has been strong leadership and the will to follow through. Without these intrinsic elements, the United States has pursued a course of action akin to "two steps forward and one step back." In sum, the United States has not adequately defined the problem or applied all the means at its disposal for solving it. We have been terribly remiss in eliminating "strategic minerals dependence" as a problem, despite being the strongest economic, military, political and technological power in the world. In many cases we have simply stuck our head in the sand in the belief that if we ignore the problem long enough, it will go away.

Although these indictments may seem contradictory to vulnerability analysis (chapter 2), they really are not. The purpose of that chapter was to present the other side of the story to those who believe that "the sky is falling." The thrust of this chapter is to be thought-provoking and to get more people thinking about the problem in a larger context in order to spur more innovative thinking, fresh ideas, and new approaches to an age-old problem. In conclusion, the words of Francis Bacon from his book *Advancement of Learning* (circa 1605) seem perhaps appropriate. In his chapter titled "Of Studies" he offered a useful maxim:

> *Read not to contradict and refute, nor to believe and take for granted, nor to find talk and discourse, but to weigh and consider.*

Appendix A
World and US Minerals Data

Table A-1. US and World Reserve and Cumulative Demand to the Year 2000 (All Data Rounded to Two Significant Figures)

	US Reserve	*US 1983-2000 Demand*	*World Reserve*	*World 1983-2000 Demand*
Antimony	90,000 st	440,000 st	4 million st	1 million st
Arsenic	50 kt	310 kt	1 million Mt	590 kt
Asbestos	4 Mt	5.6 Mt	110 Mt	110 Mt
Barite	11 million st	60 million st	160 million st	150 million st
Bauxite (Contained Al)	8 Mt	100 Mt	4.4 Gt	400 Gt
Beryllium	28,000 st	6,900 st	420,000 st	12,000 st
Bismuth	20 million lb	43 million lb	200 million lb	170 million lb
Boron (Bo)	120 million st	7.8 million st	360 million st	22 million st
Bromine	25 billion lb	5.9 billion lb	adequate	15 billion lb
Cadmium	90 kt	75 kt	560 kt	350 kt
Cesium	0	300 st	110,000 st	520 st
Chromium	0	7.9 million st	360 million st	74 million st
Cobalt	0	390 million lb	8 billion lb	1.2 billion lb
Columbium	0	200 million lb	9 billion lb	870 million lb
Copper	57 mt	31 Mt	340 Mt	170 Mt
Corundum	0	17,000 st	7 million st	540,000 st
Diamonds, industrial stones	0	60 million ct	600 million ct	460 million ct
Diatomite	250 million st	10 million st	800 million st	29 million st
Feldspar	adequate	13 million st	adequate	72 million st
Fluorspar	36 million st	12 million st	850 million st	110 million st
Gallium	2 million kg	290 kg	110 million kg	740,000 kg
Garnet	5 million st	550,000 st	8.1 million st	860,000 st
Germanium	450 kg	990 kg	adequate	2.8 million kg
Gold	80 million tr oz	52 million tr oz	1.3 billion tr oz	670 million tr oz
Graphite crystalline flake	0	400,000 st	15 million st	7.8 million st
Gypsum	800 million st	480 million st	2.6 billion st	1.9 billion st
Hafnium	80,000 st	1,400 st	460,000 st	2,700 st
Helium	240 billion cu ft	29 billion cu ft	240 billion cu ft	41 billion cu ft
Indium	7 million tr oz	14 million tr oz	54 million tr oz	40 million tr oz
Iodine	550 million lb	150 million lb	5.9 billion lb	660 million lb
Iron ore (contained Fe)	3.7 billion st	900 million st	72 billion st	9.9 billion st
Kyanite	adequate	2.1 million st	adequate	11 million st
Lead	21 Mt	12 Mt	95 Mt	61 Mt
Lithium	400,000 st	64,000 st	2.1 million st	180,000 st

Mining Engineering, April 1986, p. 246.

Table A-1—*Continued*

	US Reserve	*US 1983-2000 Demand*	*World Reserve*	*World 1983-2000 Demand*
Magnesite (contained Mg)	10 million st	14 million st	2.8 billion st	110 million st
Manganese	0	14 million st	1 billion st	170 million st
Mercury	140,000 fl	700,000 fl	4 million fl	3.7 million fl
Mica (sheet)	0	29 million lb	adequate	190 million lb
Molybdenum	6 billion lb	1.1 billion lb	12 billion lb	3.6 billion lb
Nickel	300,000 st	3.8 million st	58 million st	18 million st
Peat	700 million st	27 million st	adequate	8.5 billion st
Perlite	50 million st	11 million st	700 million st	35 million st
Phosphate	1.4 billion Gt	700 Mt	14 Gt	3.2 Gt
Platinum-group metals	1 million tr oz	34 million tr oz	1 billion tr oz	130 million tr oz
Potash (K_2O equivalent)	95 Mt	110 Mt	9.1 Gt	590 Mt
Pumice	adequate	13 million st	adequate	260 million st
Rare Earths (REO) and Yttrium (Y_2O_2)	4.9 Mt	460 kt	45 Mt	810 kt
Rhenium	2 million lb	190,000 lb	6.4 million lb	340,000 lb
Rubidium	0	49,000 lb	4.4 million lb	91,000 lb
Salt	adequate	833 million st	adequate	4.2 billion st
Sand and gravel	adequate	14 billion st	adequate	adequate
Scandium	230 t	770 kg	770 t	14 t
Selenium	12 kt	10 kt	80 kt	28 kt
Silicon alloys	adequate	10 million st	adequate	61 million st
Silver	920 million tr oz	1.9 billion tr oz	7.9 billion tr oz	5.4 billion tr oz
Soda ash	26 billion st	130 million st	26 billion st	720 million st
Stone	adequate	18 billion st	adequate	adequate
Strontium	0	450,000 st	7.5 million st	1.2 million st
Sulfur	160 Mt	250 Mt	1.3 Gt	1.3 Gt
Talc	150 million st	28 million st	350 million st	210 million st
Tantalum	0	27 million lb	60 million lb	43 million lb
Tellurium	3.7 kt	2.5 kt	22 kt	4.5 kt
Thallium	70,000 lb	48,000 lb	830,000 lb	450,000 lb
Thorium	220 kt	770 kt	1.1 Mt	6.5 kt
Tin	20 kt	700 kt	3.1 Mt	3.9 Mt
Titanium	8.1 million st	11 million st	190 million st	42 million st
Tungsten	150 kt	230 kt	2.8 Mt	970 kt
Vanadium	190,000 st	130,000 st	4.8 million st	870,000 st
Vermiculite	25 million st	6.3 million st	50 million st	11 million st
Zinc	22 Mt	19 Mt	170 Mt	130 Mt
Zirconium	4 million st	1.3 million st	23 million st	5.2 million st

Table A-2. USSR Exports of Platinum-Group Metals to the West[a] (Thousand Troy Ounces)

	1980	*1981*	*1982*	*1983*	*1984*	*1985*[b]	*1986*[c]
Total	1,686	1,652	1,571	1,595	1,837	1,508	1,890
Platinum	403	331	319	264	228	238	270
United States	29	29	14	18	20	22	45[d]
Western Europe	214	133	109	128	89	110	NA[e]
Japan	160	169	196	118	119	106	NA
Palladium	1,214	1,285	1,217	1,289	1,565	1,222	1,540
United States	340	324	378	389	495	273	194
Western Europe	383	428	209	168	221	274	NA
Japan	491	533	630	732	849	675	NA
Other	69	36	35	42	44	48	80
United States	51	20	11	12	11	17	27[d]
Western Europe	8	6	7	3	8	10	NA
Japan	10	10	17	27	25	21	NA

[a]. According to US Bureau of Mines figures for 1984-87, 9 percent of PGM Imports to the US came from the USSR. The USSR does not publish platinum-group metal trade statistics. The figures are derived mainly from partner country trade statistics published by the Organization for Economic Cooperation and Development.
[b]. Preliminary.
[c]. Estimated by Bureau of Mines analysts.
[d]. From Department of Commerce.
[e]. N. A. means not available.

Table A-3. USSR Exports of Manganese Concentrate (Thousand Metric Tons)

	1980	*1981*	*1982*	*1983*	*1984*	*1985*	*1986*
Total	1,255	1,194	1,144	1,079	1,081	1,126	1,101
United States	0	0	0	0	0	0	0
Western Europe	0	0	0	0	0	0	0
Japan	0	0	0	0	0	0	0
Eastern Europe	1,183	1,150	1,096	1,032	1,026	1,031	1,000
Other[a]	72	44	48	47	55	95	101

Vneshnyaya torgovlya SSSR (annual issues).
[a]. North Korea and other unspecified countries.

Table A-4. USSR Exports of Chromium Ore (Thousand Metric Tons)

	1980	1981	1982	1983	1984	1985	1986
Total	576	576	561	496	442	471	474
United States	99	76	11	0	0	0	20
Western Europe	33	38	21	20	0	0	0
Japan	0	30	76	78	30	58	10
Eastern Europe	371	405	427	371	383	385	383
Other[a]	73	27	26	27	29	28	61

Vneshnyaya torgovlya SSSR (annual issues).

[a]. North Korea and other specified countries.

Table A-5. Percentage Share of the World's Production for 25 Minerals Among the 5 Major Mineral-Producing Countries

	United States	Canada	Australia	Republic of South Africa	USSR	Total
Platinum-group metals	W	4	—	50	43	97
Vanadium	W	9	—	57	29	95
Nickel	W	30	9	4	23	66
Potash	5	26	—	—	34	65
Chromium	W	—	—	35	28	63
Manganese	—	—	8	14	40	62
Phosphate	30	—	—	2	26	58
Diamond	—	—	31	10	13	54
Molybdenum	40	13	—	—	—	53
Iron ore	6	4	11	2	27	50
Bauxite	1	—	37	—	5	43
Tantalum	—	10	32	—	—	42
Lead	11	11	14	3	—	39
Copper	17	8	3	—	8	36
Sulfur	19	12	—	3	—	34
Zinc	4	20	10	—	—	34
Silver	10	10	—	—	11	31
Tungsten	1	—	3	—	22	26
Cadmium	9	9	5	—	—	23
Asbestos	1	18	—	3	—	22
Tin	—	2	5	—	14	21
Fluorspar	1	—	—	7	11	19
Cobalt	—	6	4	—	6	16
Columbium	—	15	—	—	—	15
Antimony	W	—	—	12	—	12

W (withheld)

Data based on US Bureau of Mines 1988 production estimates. Due to fluctuations in metals markets, these one-year production estimates may not represent each country's typical market share.

Table A-6. National Defense Stockpile of Strategic and Critical Materials[a] (Inventory as of 31 March 1988)

			Inventory		Inventory Quantity	
	Unit	Goal	Quantity	Value (Millions)	Excess	Deficit
Aluminum metal group	ST Al	7,150,000	4,278,914	$ 829.0		*2,871,086
Alumina	ST	0	0	—		
Aluminum	ST	700,000	2,082	4.5		697,918
Bauxite, metal grade, Jamaica type	LDT	21,000,000	12,457,740[b]	560.6		8,542,260
Bauxite, metal grade, Surinam type	LDT	6,100,000	5,299,597	263.9		800,403
Aluminum oxide, abrasive grain group	ST Ab	638,000	259,043	128.6		*378,957
Aluminum oxide, abrasive grain	ST	0	50,904	63.6	50,904	
Aluminum oxide, fused, crude	ST	0	249,867	65.0	249,867	
Bauxite, abrasive grade	LCT	1,000,000	0	—		1,000,000
Antimony	ST	36,000	36,006	81.0	6	

Department of Defense, *Strategic and Critical Materials Report to the Congress*, September 1988, p. 35.

* Equivalent quantity.

[a]. Contains material held in Defense Production Act inventory, some of which is held against National Defense Stockpile (NDS) Goals. In addition to the materials listed above, the NDS inventory also contains asbestos (crocidolite), celestite, kyanite, mica (muscovite block, stained and lower), rare earths, and talc (ground), valued at $1.8 million. These materials have not been determined to be strategic and critical and, consequently, do not have NDS goals.

[b]. Bauxite, Metal Grade, Jamaica Type: Includes material in the physical custody of GSA, title to which is scheduled to be transferred to the Stockpile during Fiscal Years 1988–1990.

[c]. Reductions in the Chromite, Metallurgical Grade Ore and Manganese, Metallurgical Grade Ore inventories reflect outshipments of material for upgrading to High-Carbon Ferrochromium and High-Carbon Ferromanganese, respectively, under the President's Ferroalloy Upgrading Program and P.L. 99-661.

[d]. The President's Ferroalloy Upgrading Program and P.L. 99-661 have provided for increases in the inventory in excess of the 1 October 1984, goals for these upgraded forms. Consequently, no excess quantity is listed.

Table A-6—*Continued*

			Inventory		Inventory Quantity	
	Unit	*Goal*	*Quantity*	*Value (Millions)*	*Excess*	*Deficit*
Asbestos, amosite	ST	17,000	34,006	23.8	17,006	
Asbestos, chrysotile	ST	3,000	10,703	19.5	7,703	
Bauxite, refractory	LCT	1,400,000	274,229	63.9		1,125,771
Beryllium Metal Group	LB Be	2,440,000	2,170,885	258.2		*269,115
Beryl ore (11% BeO)	LB	36,000,000	35,712,000	16.0		288,000
Beryllium copper master alloy	LB	15,800,000	14,773,731	111.7		1,026,269
Beryllium metal	LB	800,000	580,000	130.5		220,000
Bismuth	LB	2,200,000	2,081,298	11.9		118,702
Cadmium	LB	11,700,000	6,328,809	53.1		5,371,191
Chromium, chem. & metallurgical group	ST Cr	1,353,000	1,263,932	1,100.6		*67,805
Chromite, chemical grade ore	SDT	675,000	242,414	12.1		432,586
Chromite, metallurgical grade ore	SDT	3,200,000	1,950,982[c]	209.7		1,249,018
Chromium, ferro, high carbon	ST	185,000	537,621	385.8	[d]	
Chromium, ferro, low carbon	ST	75,000	318,942	418.1	243,942	
Chromium, ferro, silicon	ST	90,000	58,357	48.4		31,643
Chromium metal	ST	20,000	3,763	26.5		16,237
Chromite, refractory grade ore	SDT	850,000	391,414	39.1		458,586
Cobalt	LB Co	85,400,000	53,105,165	363.8		32,294,835

Table A-6—*Continued*

			Inventory		*Inventory Quantity*	
	Unit	*Goal*	*Quantity*	*Value (Millions)*	*Excess*	*Deficit*
Columbium group	LB Cb	4,850,000	2,713,469	14.0		*2,136,531
Columbium carbide powder	LB Cb	100,000	21,372	.6		78,628
Columbium concentrates	LB Cb	5,600,000	2,019,218	6.5		3,580,782
Columbium, ferro	LB Cb	0	930,911	5.3	930,911	
Columbium metal	LB Cb	0	44,851	1.6	44,851	
Copper	ST	1,000,000	29,047	75.2		970,953
Cordage fibers, abaca	LB	155,000,000	0	—		155,000,000
Cordage fibers, sisal	LB	60,000,000	0	—		60,000,000
Diamond, industrial, group	KT	29,700,000	29,791,306	306.8	91,306	
Diamond dies, small	PC	60,000	25,473	1.2		34,527
Diamond, industrial, crushing bort	KT	22,000,000	22,001,344	38.5	1,344	
Diamond, industrial, stones	KT	7,700,000	7,777,225	267.1	77,225	
Fluorspar, acid grade	SDT	1,400,000	895,983	165.8		504,017
Fluorspar, metallurgical grade	SDT	1,700,000	411,738	51.5		1,288,262
Germanium	KG	146,000	4,858	5.2		141,142
Graphite, natural, Ceylon, Amorphous lump	ST	6,300	5,497	10.7		803
Graphite, natural, Malagasy, crystalline	ST	20,000	17,835	53.5		2,165

Table A-6—*Continued*

	Unit	Goal	Inventory		Inventory Quantity	
			Quantity	Value (Millions)	Excess	Deficit
Graphite, natural, other than Ceylon & Malagasy	ST	2,800	2,803	2.0	3	
Iodine	LB	5,800,000	6,572,016	52.9	772,016	
Jewel bearings	PC	120,000,000	75,504,579	85.5		44,695,179
Lead	ST	1,100,000	601,018	408.7		498,982
Manganese, bat. grade group	SDT	87,000	205,665	19.0	*118,665	
Manganese, bat. grade, natural ore	SDT	62,000	202,654	14.8	140,654	
Manganese, bat. grade, synthetic dioxide	SDT	25,000	3,011	4.2		21,989
Manganese, chem. & metallurgical group	ST Mn	1,500,000	1,917,028	500.8	*417,028	
Manganese ore, chemical grade	SDT	170,000	171,806	14.1	1,806	
Manganese ore, metallurgical grade	SDT	2,700,000	2,970,393	127.0	270,393	
Manganese ferro, high carbon	ST	439,000	757,198	303.6	[d]	
Manganese ferro, low carbon	ST	0	0	—	—	
Manganese ferro, medium carbon	ST	0	29,057	20.6	29,057	
Manganese ferro, silicon	ST	0	23,574	11.1	23,574	
Manganese metal, electrolytic	ST	0	14,172	.24.4	14,172	
Mercury	FL	10,500	165,526	57.9	155,026	

Table A-6—*Continued*

	Unit	*Goal*	*Inventory*		*Inventory Quantity*	
			Quantity	*Value (Millions)*	*Excess*	*Deficit*
Mica, Muscovite block, stained & better	LB	6,200,000	5,214,461	27.8		985,539
Mica, Muscovite film, 1st & 2nd qualities	LB	90,000	1,176,377	13.8	1,086,377	
Mica, Muscovite splittings	LB	12,630,000	14,419,034	21.6	1,789,034	
Mica, Phlogopite block	LB	210,000	130,745	.7		79,255
Mica, Phlogopite splittings	LB	930,000	1,518,930	3.0	588,930	
Molybdenum group	LB Mo	0	0	—	—	
Molybdenum disulphide	LB Mo	0	0	—	—	
Molybdenum, ferro	LB Mo	0	0	—	—	
Morphine sulfate and related analgesics	AMA LB	130,000	71,303	24.2		*58,697
Crude	AMA LB	0	31,795	4.7	31,795	
Refined	AMA LB	130,000	39,508	19.5		90,492
Natural insulation fibers	LB	1,500,000	0	—		1,500,000
Nickel	ST	200,000	37,215	412.2		162,785
Platinum-group metals, iridium	Tr Oz	98,000	29,590	9.4		68,410
Platinum-group metals, palladium	Tr Oz	3,000,000	1,264,601	154.0		1,735,399
Platinum-group metals, platinum	Tr Oz	1,310,000	452,641	220.4		857,359
Pyrethrum	LB	500,000	0	—		500,000
Quartz crystals	LB	600,000	1,848,655	1i.1	1,248,655	

Table A-6—*Continued*

	Unit	Goal	Inventory		Inventory Quantity	
			Quantity	Value (Millions)	Excess	Deficit
Quinidine	Av Oz	10,100,000	2,473,050	10.4		7,626,950
Quinine	Av Oz	4,500,000	3,246,164	7.8		1,253,836
Ricinoleic/sebacic acid products	LB	8,800,000	5,009,697	9.7		3,790,303
Rubber	MT	864,000	127,313	154.1		736,687
Rutile	SDT	106,000	39,186	14.3		66,814
Sapphire and ruby	KT	0	16,305,502	.2	16,305,502	
Silicon carbide, crude	ST	29,000	72,800	32.8	43,950	
Silver, fine	Tr Oz	0	111,381,707	707.7	111,381,707	
Talc, steatite block & lump	ST	28	1,081	.4	1,053	
Tantalum group	LB Ta	7,160,000	2,647,748	163.3		*4,512,252
Tantalum carbide powder	LB Ta	0	28,688	4.7	28,688	
Tantalum metal	LB Ta	0	201,133	44.2	201,133	
Tantalum minerals	LB Ta	8,400,000	2,837,943	114.4		5,562,057
Thorium nitrate	LB	600,000	7,121,812	19.6	6,521,812	
Tin	MT	42,700	177,053	1,245.2	134,353	
Titanium sponge	ST	195,000	36,831	295.5		158,169
Tungsten group	LB W	50,666,000	71,809,018	303.6	*21,143,018	
Tungsten carbide powder	LB W	2,000,000	2,032,942	20.3	32,942	

Table A-6—*Continued*

	Unit	Goal	Inventory Quantity	Inventory Value (Millions)	Inventory Quantity Excess	Inventory Quantity Deficit
Tungsten, ferro	LB W	0	2,025,361	8.7	2,025,361	
Tungsten metal powder	LB W	1,600,000	1,898,831	17.2	298,831	
Tungsten ores & concentrates	LB W	55,450,000	77,381,762	257.4	21,931,767	
Vanadium group	ST V	8,700	721	8.6		*7,979
Vanadium, ferro	ST V	1,000	0	—		1,000
Vanadium pentoxide	ST V	7,700	721	8.6		6,979
Vegetable tannin extract, chestnut	LT	5,000	12,355	8.4	7,355	
Vegetable tannin extract, quebracho	LT	28,000	123,805	85.1	95,805	
Vegetable tannin extract, wattle	LT	15,000	14,998	10.6		2
Zinc	ST	1,425,000	378,316	386.9		1,046,684
TOTAL VALUE OF INVENTORY				$9,144.4		

Table A-7. Principal World Supply Sources in 1984

	Production (% of Total)	*Reserves (% of Total)*
	Chromite	
USSR	35.3	12.2
South Africa	31.3	78.4
Albania	9.1	0.6
Turkey	5.3	0.4
Zimbabwe	5.2	1.6
India	4.6	1.3
Total	3.3 million s.t. (Cr content)	359 million s.t. (Cr content)
	Ferrochromium	
South Africa	31.1	
USSR	14.8	
Japan	11.4	
Zimbabwe	6.0	
Brazil	4.4	
Sweden	4.4	
China	4.1	
Others[1]	23.8	
Total	3.1 million s.t. (gross weight)	
	Manganese Ore	
USSR	36.2	36.5
South Africa	14.6	40.7
Brazil	11.8	2.1
Gabon	11.7	11.0
Australia	9.9	7.5
Total	9.2 million s.t. (Mn content)	1,000 million s.t. (Mn content)

U.S. Bureau of Mines, *South Africa and Critical Materials, Open File Report 76-86.*

NA Not available.
s.t. = short tons
[1] Distributed among 16 countries.
[2] Distributed among 26 countries.
[e] Estimated.
lb. = pounds
tr. oz. = troy ounces.

Table A-7—*Continued*

	Production (% of Total)	*Reserves (% of Total)*
	Ferromanganese	
USSR	31.4	
China	8.8	
Japan	8.7	
South Africa	6.9	
France	4.9	
Norway	4.5	
Others[2]	34.8	
Total	6.1 million s.t. (gross weight)	
	Platinum-Group Metals	
USSR	52.5	19.0
South Africa	41.1	79.0
Canada	4.9	0.8
Total	7.05 million tr.oz.	1,000 million tr. oz.
	Cobalt	
Zaire	54.6	37.5
Zambia	14.9	10.0
USSR	8.5	3.8
Canada	6.4	1.2
Cuba	4.9	28.8
Australia	4.1	0.6
Finland	2.9	0.6
New Caledonia	0.8	6.2
Philippines	0.4	3.8
Indonesia	—	5.0
Total	68.5 million lb. (Co content)	8,000 million lb. (Co content)
	Vanadium	
South Africa	37.6	19.8
USSR	28.6	60.4
China	13.6	14.0
Finland	9.2	0.7
United States	8.9	3.8
Total	36.7 thousand s.t. (V content)	4,800 thousand s.t. (V content)

Table A-8. US Dependence on South Africa for the "Big Four" (Percent)

	US—Total Import Dependence	*Dependence on South Africa*
Chromium	77	56[a]
		47[b]
Cobalt	95	50[c]
Manganese	100	26
		30[d]
PGMs	90	41[e]

[a]. Chromite (chromium ore)
[b]. Ferrochromium
[c]. Imports from Zaire and Zambia shipped through South Africa
[d]. Manganese ferroalloys
[e]. Direct shipments only

U.S. Bureau of Mines, Department of Commerce, General Services Administration, 15 January 1987.

Table A-9. US Dependence on Imports from South Africa (1985)

	Production		South Africa (Percent of World Production)	US Imports from South Africa	US Net Import Reliance (Percent)	Imports from South Africa as a Percentage of US:	
	US	South Africa				Imports	Consumption
Chromium (chromite) st	0	3,682,000	34	301,000	77	73	56
Cobalt, st							
Total	0	NA	NA	146	95	2	2
Transshipped by Zaire, Zambia, and Zimbabwe)	—	14,425	53	4,560	95	53	50
Manganese, st							
All forms	0	1,587,000	16	156,800	100	26	26
Battery grade	w[1]	NA	NA	1,888	56[2]	7	4[2]
Metal	w	35,081	NA	8,402	38[3]	98	38[3]

Bureau of Mines, Department of Commerce, General Services Administration; prepared 15 January 1987.

Note: Precise data on quantities transshipped through South Africa are generally not available.

[1] Withheld company proprietary data.

[2] Estimated Based on 1984 data.

[3] Estimated.

[4] All forms (UK).

[5] Less than 1 percent.

[6] Excludes Ceylon and Malagasy.

Table A-9—*Continued*

	Production		*South Africa (Percent of World Production)*	*US Imports from South Africa*	*US Net Import Reliance (Percent)*	*Imports from South Africa as a Percentage of US:*	
	US	*South Africa*				*Imports*	*Consumption*
Platinum-group metals (troy ounces)							
All forms	3,987	3,700,000	47	1,816,000	90	46	41
Iridium	0	NA	NA	12,400	NA	50	NA
Palladium	3,463	NA	NA	584,000	NA	39	NA
Platinum	524	NA	NA	1,029,000	NA	69	NA
Indirect imports[4]	—	—	—	340,000[3]	90	9[3]	8[3]
Ferroalloys, st							
Ferrochromium	110,000	950,000	29	203,500	77	61	47
Transshipped (Zimbabwe)	—	198,000	6	41,400	77	13	10
Ferromanganese	154,000	630,000	10	199,300	79	37	30
Ferronickel	w	NA	NA	24	NA	1[5]	1[5]
Ferrovanadium							
Indirect imports	w	NA	NA	1,420	38[3]	91	34[3]

Table A-9—*Continued*

	Production		South Africa (Percent of World Production)	US Imports from South Africa	US Net Import Reliance (Percent)	Imports from South Africa as a Percentage of US:	
	US	South Africa				Imports	Consumption
Aluminum metal group, st	8,858,000	185,000	1	17,565	16	1	1[5]
Antimony, st	w	8,150	13	2,933	62	15	9
Asbestos, st	0	41,730	NA	133	100	100	100
Almosite	61,400	101,000	NA	767	70	1	1
Chrysotile (Transshipped—Zimbabwe)	—	181,900	NA	161	70	1[5]	1[5]
Beryllium metal group, st (beryl ore)	5,738	5	1[5]	39	20	2	1[5]
Copper, st Total	1,219,000	276,000	3	6,562	28	1	1[5]

Table A-9—*Continued*

	Production		*South Africa (Percent of World Production)*	*US Imports from South Africa*	*US Net Import Reliance (Percent)*	*Imports from South Africa as a Percentage of US:*	
	US	*South Africa*				*Imports*	*Consumption*
Transshipped from Zimbabwe and Zaire, Zambia, and Botswana	—	1,197,000	13	44,170	28	10	3
Diamonds, cts (natural, industrial stones)	0	5,652,000	15	4,631,000	92	29	28
Fluorspar, st (acid grade)	w	342,000	19[3]	181,000	70[3]	42	29[3]
Graphite,[6] (natural)							
Total	0	0	0		100	0	0
Transshipped from Zimbabwe	—	13,200	2	699	100	1	1

Table A-9—*Continued*

	Production		*South Africa (Percent of World Production)*	*US Imports from South Africa*	*US Net Import Reliance (Percent)*	*Imports from South Africa as a Percentage of US:*	
	US	*South Africa*				*Imports*	*Consumption*
Nickel, st							
Total	6,127	27,600	3	7,160	72	5	3
Transshipped from Zimbabwe and Botswana	—	31,000	4	13,400	72	8	6
Pyrethrum, lbs	0	NA	NA	3,060	100	2	2
Rutile	w	261,000	35	82,000	62	39	24
Tantalum, st							
Total	0	2	1[5]	0	89	0	0
Transhipped from Zaire Zimbabwe, and Botswana	—	65	21	29	89	8	7

Table A-9—*Continued*

	Production		*South Africa (Percent of World Production)*	*US Imports from South Africa*	*US Net Import Reliance (Percent)*	*Imports from South Africa as a Percentage of US:*	
	US	*South Africa*				*Imports*	*Consumption*
Tin, st							
Total	w	3,506	2	151	72	1[5]	1[5]
Transshipped from Namibia and Zimbabwe	—	5,097	2	154	72	1[5]	1[5]
Vanadium, st							
Total	w	15,449	46	484	46[3]	59	27[2]
Indirect imports	—	—	—	38	46[3]	5	2[3]
Zinc, st							
Total	277,000	139,550	2	4,600	74	1	1[5]
Transshipped from Zaire and Zimbabwe	—	121,250	2	13,300	74	2	1

Table A-10. Selected Minerals: US Production and Consumer Stocks

	1982	*1983*	*1984*	*1985*	*1986*[e]	*1986 Stocks Consumption Ratio*
Aluminum (thousand mt)[1]	2,182	2,265	2,653	2,341	2,250	.43
Antimony (st)[2]	5,973	3,935	6,895	6,003	6,500	.54
Chromium[3] (thousand st)	209	175	127	122	134	.29
Cobalt (st)	3,733	4,305	5,411	5,977	6,177	.82
Manganese (thousand st)						
ore	751	617	582	589	590	1.11
ferromanganese	243	195	159	100	100	.23
Platinum group (thousand tr. oz.)	1,107	943	1,319	1,191	1,000	.27
Rutile[4]	176	130	102	116	106	.33
Titanium[5] (st)	3,350	3,136	3,147	4,755	3,600	.18
Tungsten[6] (mt)	1,819	1,132	1,005	1,137	900	.17
Vanadium (thousand lbs.)	11,096	9,674	7,540	6,418	5,463	.54

e. Estimated
1 Metric tons
2 Short tons
3 Consumer stocks
4 Titanium—containing ore
5 Sponge metal
6 Concentrate

Table A-11. World Chromite Production and Capacity in 1984 (Thousand Short Tons Chromium Content Unless Otherwise Specified)

	Production[e]	*Available Capacity*	*Capacity Utilized (Percent)*	*Excess Capacity*
USSR	1,152	1,152	100.0	0
South Africa	1,021	1,460	69.9	439
Albania	296	323	91.6	27
Turkey	174	240	72.5	66
Zimbabwe	171	377	45.4	206
India	149	190	78.4	41
Brazil	90	125	72.0	35
Philippines	68	170	40.0	102
Finland	46	187	24.6	141
New Caledonia	31	33	93.9	2
Iran	18	30	60.0	12
Cuba	11	29	38.0	18
Greece	11	17	64.7	6
Madagascar	9	42	21.0	33
Sudan	9	10	90.0	1
Other	8	8	100.0	0
Total	3,264	4,393		1,129

[e]. Estimated.

US, Bureau of Mines, *South African and Critical Materials*. July 1986.

Table A-12. World Cobalt Production and Capacity[1] in 1984 (Thousand Pounds Cobalt Content Unless Otherwise Specified)

	Production[e]	*Available Capacity*	*Capacity Utilized (Percent)*	*Excess Capacity*
Zaire	37,400	37,000	100.0	0
Zambia	10,180	10,000	100.0	0
USSR	5,800	6,000	96.7	200
Canada	4,400	14,000	31.4	9,600
Cuba	3,380	6,000	56.3	2,620
Australia	2,800	4,500	62.2	1,700
Finland	2,000	3,000	66.7	1,000
Albania	800	NA	NA	NA
New Caledonia	560	700	80.0	140
Botswana	500	700	71.4	200
Philippines	280	2,800	10.0	2,520
Brazil	220	NA	NA	NA
Zimbabwe	170	200	85.0	30
South Africa	NA	500	NA	NA
Total	68,490	85,400[2]		18,010[3]

US, Bureau of Mines.

[e]. Estimated. NA (not available).

[1] Cobalt is produced as a byproduct of copper or nickel, hence, "'available capacity"' depends on full utilization of copper and nickel production capacity.

[2] World cobalt refinery capacity is 84,900 thousand pounds.

[3] Total of column does not equal total available capacity column less total production column because of "NA."

Table A-13. World Manganese Production and Capacity in 1984 (Thousand Short Tons Manganese Content Unless Otherwise Specified)

	Production	*Available Capacity*	*Capacity Utilized (Percent)*	*Excess Capacity*
USSR	3,330	3,800	87.6	470
South Africa	1,341	3,300	40.6	1,959
Gabon	1,078	1,300	82.9	222
Brazil	1,088	1,350	80.6	262
Australia	915	1,300	70.4	385
China	530	550	96.4	20
India	520	800	65.0	280
Mexico	217	300	72.3	83
Ghana	53	150	35.3	97
Other	125	250	50.0	125
Total	9,197	13,000		3,903

US, Bureau of Mines.
[e]. Estimated.

Table A-14. World PGM Production and Capacity in 1984 (Thousand Troy Ounces Unless Otherwise Specified)

	Production[e]	*Available Capacity*	*Capacity Utilized (Percent)*	*Excess Capacity*
USSR	3,700	4,000	92.5	300
South Africa	2,900	4,100	70.7	1,200
Canada	348	600	58.0	252
Japan[1]	55	80	68.8	25
United States	15	10	100.0	0
Australia	14	20	70.0	6
Colombia	10	50	20.0	40
Finland	4	20	20.0	16
Yugoslavia	3	10	30.0	7
Zimbabwe	3	10		7
Total	7,053	8,900		1,853

US, Bureau of Mines.

[e]. Estimated.

[1] Smelter-refinery recovery from ores originating in other countries.

Table A-15. World Vanadium Production and Capacity in 1984 (Thousand Tons Vanadium Content Unless Otherwise Specified)

	Production[e]	*Available Capacity*	*Capacity Utilized (Percent)*	*Excess Capacity*
Ores, concentrates, and slags[1]				
South Africa	13,798	19,000	72.6	5,202
USSR	10,500	23,000	45.6	12,500
China	5,000	7,200	69.4	2,200
Finland[2]	3,377	4,200	80.4	823
United States	1,617	9,900	16.3	8,183
Chile	—	1,200	0	1,200
Australia	—	1,000	0	1,000
Total	34,292	65,500		31,208

US, Bureau of Mines.

[e]. Estimated.

[1] Production credited to country that was the origin of the vanadiferous raw materials.

[2] Phasing out of mining operations has begun.

Table A-16. Worldwide Ferroalloy Production and Unused Capacity (Thousand Short Tons-Gross Weight)

	1984 Production	*1984 Capacity*	*Unused Capacity*
Australia	145	166	21
Brazil	753	932	179
Canada	244	350	106
China	989	1,029	40
France	776	998	222
West Germany	453	640	187
Greece	77	105	28
India	251	625	374
Italy	236	391	155
Japan	1,564	2,394	830
Mexico	258	258	0
Norway	1,057	1,291	234
Philippines	56	98	42
South Africa	1,617	1,966	349
Spain	275	542	267
Sweden	196	446	250
Turkey	61	176	115
USSR	3,643	4,122	479
Yugoslavia	335	356	21
Zimbabwe	220	323	103
Other Africa	7	68	61
Other South America	208	267	59
Other Western Europe	454	746	292
Eastern Europe	1,033	1,258	225
Total	15,395	20,038	4,643
Communist	6,000	6,765	765
Non-Communist	9,395	13,273	3,878

US, Bureau of Mines, "Minerals Facts and Problems," 1985.

Table A-17. Selected US: Strategic Mineral Production Potential

	Maximum Output	*Percent of Current Consumption*	*Lead Time (Years)*	*Necessary Price*[1]
Chromium	175,000	41	2	2.0
(tons)	235,000	54	3	2.0
Cobalt	9.7	60	2–4	3.0
(lbs.)				
Manganese	818,000	25	5	4.7–
(tons)				21.0
Platinum group	174,000	7	2	1.0
(troy ounces)	437,000	17	3–4	1.0

US, Bureau of Mines.

1. Ratio of price needed for profitable operations to current market price.

Table A-18. First-Tier Strategic Materials Supply Prospects

	Minerals	*Country Producers*[a]	*Importance*[b] *Now*	*Importance*[b] *Potential*	*Primary Constraints to Increased Availability*[c]
North America	Chromium	NA			
	Cobalt	CANADA	2	2	nickel demand, refinery limits
		United States	—	3	High cost deposits, demand for various primary metals
	Manganese	MEXICO	3	2	customer acceptance of quality
	PGM	CANADA	3	3	nickel/copper demand, refinery limits

NA—Not applicable.

[a]. UPPERCASE indicates a current producer.

[b]. Key: 1 = major
2 = medium
3 = minor
? = unknown

Based on assessment of relative production levels and contributions to Western trade.

[c]. Any expansion/development would require capital investment to a varying degree. In mining, processing and refining infrastructure.

Office of Technology Assessment, 1984.

Table A-18—*Continued*

	Minerals	*Country Producers*[a]	*Importance*[b] *Now*	*Importance*[b] *Potential*	*Primary Constraints to Increased Availability*[c]
		United States	—	3	demand for PGMs, competition
South America	Chromium	NA			
	Cobalt	Peru	—	2	processing facilities
	Manganese	BRAZIL	2	2–3	local demand
	PGM	NA			
Australia and Oceania	Chromium	PHILIPPINES	2	2	infrastructure
		Pacific rim	—	3	proof of feasibility
	Cobalt	PHILIPPINES	2	2	demand for nickel
		AUSTRALIA	2	2	demand for nickel
		New Caledonia	—	3	demand for nickel
		Papua New Guinea	—	2	demand for cobalt/ chromium
	Manganese	AUSTRALIA	2	1–2	hauling equipment
	PGM	Pacific rim	—	?	proof of feasibility
Eurasia	Chromium	FINLAND	3	3	possible resource limits
		ALBANIA	2	1–2	unknown
		GREECE	3	3	resource limits
		TURKEY	3	3	Improved knowledge of resources and technology
		INDIA	3	3	resource limits, infrastructure, local demand
	Cobalt	FINLAND	2–3	2–3	possible resource limits
	Manganese	INDIA	3	3	resource limits, infrastructure, local demand
	PGM	NA			
Africa	Chromium	SOUTH AFRICA	1	1	transportation
		ZIMBABWE	2	1	transportation
		MADAGASCAR	3	2–3	seasonal operation, infrastructure
	Cobalt	ZAIRE	1	1	processing, refinery limits
		ZAMBIA	1	1	processing, refinery limits
		Morocco	—	3	resource evaluation
		BOTSWANA	3	2–3	transportation

Table A-18—*Continued*

	Minerals	*Country Producers*[a]	*Importance*[b] *Now*	*Potential*	*Primary Constraints to Increased Availability*[c]
	Manganese	SOUTH AFRICA	1	1	transportation
		GABON	2	2	transportation
	PGM	SOUTH AFRICA	1	1	refinery limits
Eastern Bloc	Chromium	SOVIET UNION	2	2–3	unknown
	Cobalt	SOVIET UNION	3	3	unknown
		CUBA	3	2	unknown
	Manganese	SOVIET UNION	3	3	unknown
	PGM	SOVIET UNION	1	1	unknown

Table A-19. US Recycling Effort (1988)

	Recycled scrap	*Percent of consumption*
Aluminum (million tons)	1.95	18
Chromium[a] (thousand tons)	93	25
Cobalt (tons)	1,500	19
Columbium (thousand tons)	negl.	negl.
Manganese[a] (thousand tons)	73	11
Platinum Group[a] (million tr. oz)	1.0[b]	27
Tantalum (thousand lbs.)	300	10
Tungsten (tons)	2,300	25

US, Bureau of Mines, *Mineral Commodity Summaries 1989*.

[a]. 1986 figures.

[b]. In 1988, the quantity of toll-refined secondary production was 1.5 million ounces.

Table A-20. Selected Opportunities for Increased Recycling of Chromium, Cobalt, and Platinum Group Metals

Current Level of Recycling	*Key Recovery Opportunities*	*Barriers to Increased Recycling*		
		Technical	*Economic*	*Institutional*
Superalloys (Co, Cr):				
An estimated 2.8 million lbs of cobalt-bearing superalloy processing scrap was lost or downgraded in 1980; 1.2 million lbs. of cobalt in obsolete scrap was not recovered. However, these figures are based on 1978 scrap use rates.	Increased recovery of obsolete scrap, reduced downgrading of high-quality industrial scrap and recovery of wastes, using advanced recycling technologies.	Concern about contaminants limits recycling in superalloy production to high-quality scrap and processes that have been certified.	Current prices encourage downgrading of superalloy scrap for use in stainless steel or nickel alloy production in which the cobalt is not needed.	Not significant once manufacturing specifications for use of recycled materials have been established, however, this can take several years.
		Experimental processes to reclaim elements separately have not been commercialized.	Certifying recycled materials for superalloy use is expensive and time-consuming.	
Petroleum hydroprocessing catalysts (Co):				
No cobalt recovery at this time; 270,000 lbs of cobalt was not recovered from spent catalysts in 1982.	Cobalt consumption in catalysts is expected to grow to 675,000 lbs in 1990, 90% of which could be recovered.	Not significant; various proprietary processes are purported to recover Cr, Co, Ni, Mo, Va and other metals for catalytic or chemical purposes.	Current prices may discourage recovery of cobalt in preference for molybdenum. However, at least two firms have proprietary processes for recovery of all elements in catalysts.	Landfilling of spent catalysts by refineries prevents possible future recovery of some catalysts.

Strategic Materials: Opportunities to Reduce U.S. Import Vulnerability, Office of Technology Assessment, May 1985.

Table A-20—*Continued*

Current Level of Recycling	*Key Recovery Opportunities*	*Barriers to Increased Recycling*		
		Technical	*Economic*	*Institutional*
	Accumulated spent catalyst residues may include 8 million lbs of cobalt and nickel.	Some technical problems, pilotscale recovery process is under investigation by a private firm.	Reprocessing of residue is not profitable at current prices.	Not significant; residues are stored on site of major processor.
Cemented carbides (Co): 10 to 30% of cobalt used in domestic cemented carbide shipments came from recycling.	Obsolete scrap (a high proportion of industrial scrap is already recovered.)	Not significant: several firms use a Bureau of Mines process to recover cobalt binders. One firm produces highly refined, pure cobalt powders from scrap that could potentially be used in other premium uses.	Recovery of obsolete scrap from dispersed uses may not be economic at current prices.	Lack of effective scrap collection programs by industrial users. Practical limits for postconsumer recovery vary by industry. However, some users (such as the oil industry) return 80% of their scrap for recycling; the coal mining industry, by contrast, only returns 15% (out of a practical limit of 50%).

Table A-21. Selected Pseudo-Strategic Materials

ANTIMONY

(Data in short tons of antimony content, unless noted)

1. **Domestic production and use:** Primary antimony metal and oxide were produced by nine companies operating nine plants utilizing both foreign and domestic material. Two of these plants were in Texas and one each in Idaho, Maryland, Montana, Nebraska, New Jersey, Ohio, and Tennessee. Antimony continued to be recovered as a byproduct from smelting of domestic lead ores. The estimated value of total primary antimony metal and oxide production in 1988 was $59 million. Antimony was consumed at about 250 plants in states east of the Mississippi River, accounting for approximately 90% of reported primary consumption. The principal uses of primary antimony were flame retardants, 65%; transportation, including batteries, 11%; chemicals, 9%; ceramics and glass, 5%; and other, 10%.

2. **Salient statistics—United States:**

	1984	1985	1986	1987	1988[e]
Production: Mine, silver-copper, lead, and stibnite ores	Negligible				
Refinery: Primary plants[1]	17,639	16,449	17,978	20,704	21,000
Secondary plants	14,823	15,030	15,522	16,647	18,000
Imports for consumption	23,089	20,694	25,401	26,729	32,000
Exports of metal, alloys, and oxide	991	1,247	1,175	1,653	2,000
Shipments from Govt. stockpile excesses	69	928	460	1,381	—
Consumption: Reported, primary antimony	12,465	11,697	10,952	11,086	11,500
Apparent[2]	33,261	35,076	38,259	40,965	46,000
Price, average, cents per pound[3]	151.2	131.1	121.9	110.6	107.0
Stocks, yearend	6,895	6,040	6,131	6,835	7,000
Employment, plant[e]	130	130	120	120	110
Net import reliance[4] as a percent of apparent consumption	58	63	64	63	65

3. **Recycling:** Approximately 16,200 tons of antimony was recovered from old scrap. Approximately 95% of the total secondary antimony was recovered as antimonial lead, most of which was consumed by the battery industry.

4. **Import sources (1984–87):** Metal: China, 66%; Mexico, 15%; Hong Kong, 10%; Bolivia, 2%; other, 7%.
   Ore and concentrate: Mexico, 33%; China, 18%; Bolivia, 14%; Guatemala, 10%; other, 25%.
   Oxide: Republic of South Africa, 38%; China, 29%; France, 9%; Bolivia, 6%; other, 18%.
   Total: China, 37%; Republic of South Africa, 19%; Mexico, 12%; Bolivia, 7%; other, 25%.

5. **Tariff:** See Tariff Rates.

Item	Number	Most favored nation (MFN) 1/1/89	Non-MFN 1/1/89
Ore and concentrates	2617.10.0000	Free	Free
Antimony and articles thereof, including waste and scrap	8110.00.0000	Free	4.4¢/kg
Antimony oxide	2825.80.0000	Free	4.4¢/kg

6. **Depletion allowance:** 22% (Domestic), 14% (Foreign).

US, Bureau of Mines, *Mineral Commodity Summaries*, 1989.

[e] Estimated. W = Withheld to avoid disclosing company proprietary data.

[1] Includes antimony recovered from domestic and foreign ores.

[2] Domestic mine production plus secondary production from old scrap plus net import reliance (see footnote 4).

[3] New York dealer price for 99.5% to 99.6% metal, c.i.f. U.S. ports.

[4] Defined as imports − exports + adjustments for Government and industry stock changes.

[5] See page 186 for definitions.

[6] Excludes US production.

Table A-21. ANTIMONY—*Continued*

7. **Government stockpile:**

Stockpile Status—9-30-88

Material	*Goal*	*Total inventory*	*Excess to goal authorized for disposal*	*Sales, 9 months*
Antimony	36,000	35,999	—	—

8. **Events, trends, and issues:** A smelter in Cleveland, OH, that produced antimony oxide was purchased and disassembled during 1988. The equipment from Cleveland was reportedly shipped to the new owner's smelter at Laredo, TX, where it was reinstalled to increase refinery capacity needed to process the recent increase in receipts of antimony concentrates from the new owner's mine in Mexico. Idaho's only antimony smelter was back in operation at the beginning of the third quarter of 1988.

   In 1988, imports for consumption increased considerably, and China was the major supplier of primary antimony, supplying more than one-third of the US total imported ore, metal, and oxide. Antimony production from domestic source materials was largely derived from recycling of lead antimony batteries. Recycling plus US mine output, which represented a minor tonnage, supplied only about one-half of domestic demand.

   The consumption of antimony oxide as a flame retardant continued as the dominant market for primary antimony. The use of antimony as a hardening agent in lead-acid batteries, which has declined over the last decade, was believed to have leveled off, because the range in the types of alloys used in automotive batteries has begun to narrow.

   The antimony dealer monthly average metal price published by Metals Week started the year at $1.13 per pound, decreased to $0.97 in June, and increased to $1.07 in August.

   Environmental and ecological problems associated with the treatment of antimony metal and ores are minimal because emissions and effluents are controlled at the processing plants.

9. **World mine production, reserves, and reserve base:**

	Mine production		**Reserves[5]**	**Reserve base[5]**
	1987	**1988[e]**		
United States	W	W	90,000	100,000
Bolivia	9,000	10,000	340,000	350,000
Mexico	3,300	4,000	200,000	250,000
South Africa, Republic of	7,500	8,000	260,000	280,000
Yugoslavia	1,390	1,400	100,000	100,000
Other Market Economy Countries	12,485	12,500	890,000	1,135,000
Centrally Planned Economies	28,200	28,500	2,750,000	2,960,000
World Total	[6]61,875	[6]64,400	4,630,000	5,175,000

10. **World resources:** US resources are mainly in Idaho, Nevada, Alaska, and Montana. Principal identified world resources, estimated at 5.6 million tons, are in China, Bolivia, the USSR, the Republic of South Africa, and Mexico. Additional antimony resources may occur in "Mississippi Valley Type" lead deposits in the Eastern United States.

11. **Substitutes:** Compounds of titanium, zinc, chromium, tin, and zirconium may be substituted for antimony chemicals in paint, pigments, frits, and enamels. Combinations of calcium, strontium, tin, copper, selenium, sulfur, and cadmium can be used as substitutes for hardening lead. Selected organic compounds and hydrated aluminum oxide are widely accepted alternative materials in flame-retardant systems.

Table A-21.—*Continued*

CHROMIUM

(Data in thousand metric tons, gross weight, unless noted)

1. **Domestic production and use:** Chromite mining on a regular basis stopped in 1961. Secondary chromium is recovered from stainless steel scrap. The United States consumed more than 10% of world chromium production in the form of imported chromite and chromium ferroalloys, metal, and chemicals. Imported chromite was consumed by two chemical firms to produce primary chromium chemicals, three metallurgical firms to produce chromium ferralloys, and six refractory firms to produce chromite-containing refractories. The chemical and metallurgical industry accounted for about 89% of US chromite consumption and the refractory industry, about 11%. Consumption of imported and domestic chromium ferroalloys, metal, and other chromium-containing materials by end use was about as follows: stainless and heat-resisting steel, 82%; full-alloy steel, 7%; superalloys, 3%; and other end uses, 8%. The value of chromium consumed as chromite and ferrochromium was about $386 million.

2. **Salient statistics—United States:**[1]

	1984	1985	1986	1987	1988[e]
Production: Mine	—	—	—	—	—
Secondary	80	85	84	97	132
Imports for consumption	315	291	349	322	414
Exports	33	38	35	9	5
Consumption: Reported	286	266	346	372	363
Apparent[2]	403	343	406	405	526
Price, chromite, yearend:					
Turkish, dollars per metric ton, Turkey	110	125	125	100	180
South African, dollars per metric ton, South Africa	52	42	42	46	56
Stocks, industry, yearend	116	111	102	109	124
Net import reliance[3] as a percent of apparent consumption	80	75	793	76	75

3. **Recycling:** In 1988, chromium contained in purchased stainless steel scrap accounted for 25% of chromium demand.

4. **Import sources (1984–87):** Chromium contained in chromite and ferrochromium: Republic of South Africa, 61%; Turkey, 13%; Zimbabwe, 9%; Yugoslavia, 4%; other, 13%.

5. **Tariff:**

Item	Number	Most favored nation (MFN) 1/1/89	Non-MFN 1/1/89
Ore and concentrate	2610.00.00	Free	Free
High-carbon ferrochromium	7202.41.00	1.9% ad val.	7.5% ad val.

6. **Depletion allowance:** 22% (Domestic), 14% (Foreign).

7. **Government stockpile:** In addition to the data shown, the stockpile contained the following nonstockpile-grade materials: 215,724 tons of metallurgical-grade chromite and 18,835 tons of chromium ferroalloys.

Stockpile Status—9-30-88

Material	*Goal*	*Total inventory*	*Average chromium content*	*Excess to goal authorized for disposal*	*Sales, 9 months*
Chromite:					
Metallurgical-grade	2,903	1,500	28.6%	—	—
Chemical-grade	612	220	28.6%	—	—
Refractory-grade					
Chromium ferroalloys:	771	355	—	—	—
High-carbon	168	487	71.4%	—	—
Low-carbon	68	272	71.4%	—	—
Ferrochromium-silicon	82	52	42.9%	—	—
Chromium metal	18	3	—	—	—

[e]. Estimated

[1] Data in thousand metric tons of contained chromium converted from thousand short tons as reported in the 1988 Mineral Commodity Summaries.

[2] Calculated total demand for chromium.

[3] Defined as imports – exports + adjustments for government and industry stock changes.

[4] See page 186 for definitions.

[5] Shipping-grade ore is deposit quantity and grade normalized to 45% Cr_2O_3 for high-chromium and high-iron chromite and 35% Cr_2O_3 for high-alumina chromite.

Table A-21. CHROMIUM—*Continued*

The stockpile conversion program whereby stockpiled chromium ores are being upgraded to ferrochromium was Presidentially mandated in 1982 and congressionally required for a 7-year period from 1987 until 1993 at a rate of no less than 53,500 short tons of ferrochromium per year. The Department of Defense converted about 138,000 short tons of chromite to high-carbon ferrochromium in 1988, the fifth year of this program.

8. **Events, trends, and issues:** Inadequate supplies of ferrochromium resulted from strong demand for stainless steel in all three major producing regions: the United States, Japan, and Europe. Inadequate ferrochromium supply resulted in increasing ferrochromium prices and delays in stainless steel shipments. The price of chromite rose in 1988 following the steady rise in ferrochromium price since mid-1987, when stainless steel production was strong enough and lasted long enough to deplete ferrochromium consumer and producer stocks and chromite producer stocks. In many countries, new ferrochromium plants and plant expansions are in progress or being planned. If all announced projects are completed in the next 3 years, a 20% increase over the current market economy production capacity of 3 million tons per year could occur. Strong demand for chromium combined with limited supply of ferrochromium resulted in increased consumption of stainless steel scrap.

   Chromium, including chromite and ferrochromium, continued to be a strategic mineral essential for the economy or defense of the United States that is unavailable in adequate quantities from reliable and secure suppliers under the Comprehensive Anti-Aparthied Act of 1986.

   Compared with that of 1987, apparent consumption increased 30%. Chromium demand changes reflect changes or anticipated changes in the steel industry, the major consumer of chromium. It is estimated that in 1989, US apparent consumption of chromium will be about 500,000 tons and domestic mine production of chromium ore will be zero. Domestic deposits are small or of low grade and are located far from consumers.

   Chromium releases into the environment are regulated by the Environmental Protection Agency. Workplace exposure is regulated by the Occupational Safety and Healthy Administration.

9. **World mine production, reserves, and reserve base:**

	Mine production		Reserves[4]	Reserve Base[4]
	1987	1988[e]	(shipping grade)[5]	
United States	—	—	—	—
Brazil	227	230	8,000	9,000
Finland	712	720	17,000	29,000
India	522	525	13,000	60,000
Philippines	172	175	14,000	29,000
South Africa, Republic of	3,787	4,000	828,000	5,700,000
Turkey	599	600	4,000	70,000
Zimbabwe	540	550	17,000	750,000
Other Market Economy Countries	316	320	17,000	23,000
Albania	830	830	6,000	20,000
USSR	3,148	3,200	102,000	102,000
Other Centrally Planned Economies	137	140	3,000	3,000
World Total (may be rounded)	10,990	11,290	1,030,000	6,800,000

10. **World resources:** World resources total about 33 billion tons of shipping-grade chromite, sufficient to meet conceivable demand for centuries. More than 99% of these resources are in southern Africa. Although the rest of the world's resources are measured in millions of tons, they are small in comparison with those of Africa. The largest US chromium resource is in the Stillwater Complex in Montana.

11. **Substitutes:** There is no substitute for chromite ore in the production of ferrochromium, chromium chemicals, or chromite refractories. Chromium-containing scrap can substitute for ferrochromium in metallurgical uses. There is no substitute for chromium in stainless steel, the major end use of chromium, nor for chromium in superalloys, the major strategic end use of chromium. Substitutes for chrome-containing alloys, chromium chemicals, and chromite refractories increase cost or sacrifice performance. Substitution for chromium-containing products could result in savings of about 60% of chromium used in alloying metals, about 15% of chromium used in chemicals, and 90% of chromite used in refractories given 5 to 10 years to develop technically acceptable substitutes and acceptance of increased cost.

Table A-21.—*Continued*

COBALT

(Data in short tons of cobalt content, unless noted)

1. **Domestic production and use:** Domestic mine production ceased at the end of 1971. Most secondary cobalt is derived from recycled superalloy or cemented carbide scrap and from spent catalysts. About 13 recyclers accounted for nearly all the cobalt recycled in superalloy scrap. There were two extra-fine cobalt powder producers, one a foreign-owned company producing powder primary material and the other a domestically controlled company producing powder from recycled materials. Nine processors were active in the production of cobalt compounds. About 90 industrial consumers were surveyed, with the largest consumption reported in Pennsylvania, Michigan, New York, and New Jersey. Superalloys used mainly in industrial and aircraft gas turbine engines accounted for about 38% of reported consumption; paint driers, 15%; magnetic alloys, 12%; catalysts, 9%; and other, 26%. Total estimated value of cobalt consumed in 1988 was $130 million.

2. **Salient statistics—United States:**

	1984	1985	1986	1987	1988[e]
Production: Mine	—	—	—	—	—
Secondary	440	449	1,319	1,255	1,500
Imports for consumption[1]	12,655	8,854	6,144	9,736	9,300
Exports	336	314	316	403	600
Shipments from Govt. stockpile excesses[2]	(99)	(3,495)	—	—	—
Consumption: Reported	6,472	6,771	7,221	7,550	7,900
Apparent[1]	8,948	7,846	8,687	8,910	9,600
Price, average annual spot for cathodes, dollars per pound	10.40	11.43	7.49	6.56	7.00
Stocks, industry, yearend	5,411	5,977	4,438	6,223	6,600
Employment, mine	—	—	—	—	—
Net import reliance[3] as a percent of apparent consumption	95	94	85	86	84

3. **Recycling:** About 1,500 tons of cobalt was recycled from purchased scrap in 1988. This represented about 19% of estimated reported consumption for the year.

4. **Import sources (1984–87):** Zaire, 36%; Zambia, 19%; Canada, 17%; Norway, 9% (originated in Canada); other, 19%.

5. **Tariff:**

Item	Number	Most favored nation (MFN) 1/1/89	Non-MFN 1/1/89
Unwrought cobalt, alloys	8105.10.30	5.5% ad val.	45% ad val.
Unwrought cobalt, other	8105.10.60	Free	Free
Chemical compounds:			
Oxides and hydroxides	2822.00.00	1.2¢/lb	20¢/lb
Sulfate	2833.29.10	1.4% ad val.	6.5% ad val.
Other:			
Cobalt chloride	2827.34.00	4.2% ad val.	30% ad val.
Cobalt carbonates	2836.99.10	4.2% ad val.	30% ad val.
Cobalt acetates	2915.23.00	4.2% ad val.	30% ad val.
Cobalt ores and concentrates	2605.00.00	Free	Free

6. **Depletion allowance:** 22% (Domestic); 14% (Foreign).

[e]. Estimated. NA Not available.

[1] Cobalt imports in 1984 included 2,706 tons of cobalt that was destined for the National Defense Stockpile, but did not officially enter the stockpile; 795 tons of 1985 imports was delivered to the stockpile. To derive meaningful apparent consumption figures, the 2,706 tons and the 795 tons were treated as if they had been put into stocks in 1984 and 1985, respectively. Moreover, the 99 tons of cobalt that officially entered the stockpile in 1984 was imported in 1983 and was not used in deriving the 1984 apparent consumption figure.

[2] Data in parentheses denote stockpile acquisitions.

[3] Defined as imports – exports + adjustments for Government and industry stock changes.

[4] See page 186 for definitions.

[5] Excludes Albania.

Table A-21. COBALT—*Continued*

7. **Government stockpile:**

Stockpile Status—9-30-88

Material	*Goal*	*Total inventory*	*Excess to goal authorized for disposal*	*Sales, 9 months*	*Purchases, 9 months*
Cobalt	42,700	26,553	—	—	—

8. **Events, trends, and issues:** Demand for cobalt in 1988 increased slightly. As in 1987, the spot price for cathode cobalt was stable, remaining between $6.75 and $7.20 per pound from January through September. This is a result of the cooperative efforts of Zaire and Zambia, the world's largest cobalt producers, who agreed to a producer price of $7.50 per pound for 1988. Import reliance is estimated at 84%, about the same level as in 1987. In 1988 management of the National Defense Stockpile was transferred from the General Services Administration to the Department of Defense.

   Surplus market conditions are expected to prevail worldwide in 1989. Most of the surplus is in producers' stocks. It is estimated that in 1989 domestic mine production will be zero and that US apparent consumption will be about 9,500 tons.

9. **World mine production, reserves, and reserve base:**

	Mine production		**Reserves**[4]	**Reserve Base**[4]
	1987	**1988**[e]		
United States	—	—	—	950,000
Australia	1,800	1,800	25,000	100,000
Canada	2,850	2,850	50,000	285,000
Finland	300	300	25,000	37,500
New Caledonia	825	830	250,000	950,000
Philippines	—	—	—	440,000
Zaire	32,000	32,000	1,500,000	2,300,000
Zambia	6,150	6,150	400,000	600,000
Other Market Economy Countries	1,248	1,250	100,000	1,300,000
Albania	650	650	NA	NA
Cuba	1,750	1,750	1,150,000	2,000,000
USSR	3,100	3,100	150,000	250,000
World Total (may be rounded)	50,673	50,680	[5]3,650,000	[5]9,200,000

10. **World resources:** The cobalt resources of the United States are estimated to be about 1.4 million tons. Most of these resources are in Minnesota, but other important occurrences are in Alaska, California, Idaho, Missouri, Montana, and Oregon. Although large, most domestic resources are in subeconomic concentrations that will not be economically usable in the foreseeable future. The identified world cobalt resources are about 12 million tons. The vast majority of these resources are in nickel-bearing laterite deposits, with most of the rest occurring in nickel-copper sulfide deposits hosted in mafic and ultramafic rocks and in the sedimentary copper deposits of Zaire and Zambia. In addition, millions of tons of hypothetical and speculative cobalt resources exist in manganese nodules and crusts on the ocean floor.

11. **Substitutes:** Nickel may be substituted satisfactorily for cobalt in superalloys. In other applications, nickel substitution can result in a loss in performance. Various potential substitutes include nickel, platinum, barium or strontium ferrite and iron in magnets; tungsten, molybdenum carbide, ceramics, and nickel in machinery; nickel and ceramics in jet engines; nickel in catalysts; and copper, chromium, and manganese in paint.

Table A-21.—*Continued*

DIAMOND (INDUSTRIAL)

(Data in million carats, unless noted)

1. **Domestic production and use:** Synthetic diamond production, imports, and exports established record-high levels. All industrial diamond produced domestically was synthetic grit and powder. The output was from three firms, one each in New Jersey, Ohio, and Utah. Four firms recovered and sold industrial diamond as the principal product. About 35 firms recovered industrial diamond in secondary operations. Major uses of all industrial diamond were machinery, 27%; mineral services, 18%; stone and ceramic products, 27%; abrasives, 16%; contract construction, 13%; transportation equipment, 6%; and other, 3%. The mineral services industry, primarily drilling, accounted for 85% of stone consumption.

2. **Salient statistics—United States:**

	1984	1985	1986	1987	1988[e]
Bort, grit, and powder and dust, natural and synthetic:					
Production: Manufacturer diamond	76.0	76.0	80.0	W	W
Secondary[e]	1.5	1.0	3.1	3.0	5.4
Imports for consumption	35.2	36.4	37.1	45.0	52.5
Exports and reexports[2]	48.0	51.6	51.2	56.8	64.7
In manufactured products[e]	0.5	0.6	0.6	0.6	0.6
Sales of Govt. stockpile excesses	—	—	—	—	—
Consumption, apparent	64.2	61.2	68.4	W	W
Price, value of imports, dollars per carat	1.33	1.27	1.25	1.28	0.88
Net import reliance[3] as a percent of apparent consumption	E	E	E	E	E
Stones, natural:					
Production: Mine	—	—	—	—	—
Secondary[e]	0.7	0.5	1.5	0.3	0.6
Imports for consumption	8.3	9.4	8.9	3.9	6.0
Exports and reexports[2]	3.8	3.8	4.1	3.0	2.5
Sales from Govt. stockpile excesses	[4]1.5	[4]1.5	[4]2.0	[4]2.0	1.8
Consumption, apparent	5.2	6.1	6.3	1.2	5.9
Price, value of imports, dollars per carat	8.03	8.52	7.23	10.86	7.80
Net import reliance[3] as a percent of apparent consumption	87	92	76	75	90

3. **Recycling:** About 6.0 million carats was salvaged in secondary production from salvage stone, sludge, and swarf.

4. **Import sources (1984–87):** Bort, grit, and powder and dust, natural and synthetic: Ireland, 66%; Japan, 10%; United Kingdom, 4%; Belgium-Luxembourg, 3%; other, 17%.
   Stone, natural: Republic of South Africa, 38%; United Kingdom, 19%; Ireland, 13%; Zaire, 10%; other, 20%.

[e]. Estimated. E Net exporter. W Withheld to avoid disclosing company proprietary data.
[1] Industry stocks and employment were unknown.
[2] Includes diamonds in manufactured abrasive products.
[3] Defined as imports − exports + adjustments for Government and industry stock changes.
[4] Diamond stones from the Government stockpile were apparently exported as near gem quality diamond; consequently, they were excluded from calculations.
[5] Natural industrial diamond only.
[6] See page 186 for definitions.

Table A-21. DIAMOND (INDUSTRIAL)—*Continued*

5. **Tariff:**

Item	Number	Most favored nation (MFN) 1/1/89	Non-MFN 1/1/89
Miners' diamond, carbonados	7102.10	Free	Free
Other	7102.20	Free	Free
Industrial diamond, natural advanced	7102.30	4.9%	30%
Industrial diamond, natural not advanced	7102.40	Free	Free
Industrial diamond, other	7102.29	Free	Free
Dust, grit, or powder, natural, synthetic	7105.10	Free	Free

6. **Depletion allowance:** 14% (Domestic); 14% (Foreign).

7. **Government stockpile:** Excess diamond stones were used as payment materials for the ferroalloys upgrading program.

Stockpile Status—9-30-88

Material	*Goal*	*Total inventory*	*Excess to goal authorized for disposal*	*Sales, 9 months*
Crushing bort	22.0	22.0	—	—
Industrial stones	7.7	7.7	—	1.8

8. **Events, trends, and issues:** The industrial diamond industry experienced another robust year with apparent US consumption estimated to have increased 3%. Total US sales, including exports, were up an estimated 17%. It is estimated that the industrial diamond worldwide market performed similarly to the US market. Companies in Ireland, the Republic of South Africa, Sweden, and the United States increased production capacity for synthetic industrial diamond during 1988.

9. **World mine production, reserves, and reserve base:**

	Mine production 1987	Mine production 1988[e]	Reserves[e6]	Reserve base[e6]
United States	—	—	—	Unknown
Australia	16.7	16.9	500	900
Botswana	3.8	3.8	125	200
Brazil	0.3	0.4	5	15
South Africa, Republic of	5.0	5.2	70	150
Zaire	18.7	18.7	150	350
Other Market Economy Countries	1.3	1.2	40	65
China	0.8	0.6	10	20
USSR	7.0	7.1	80	200
World Total	53.6	53.9	980	1,900

10. **World resources:** The potential to discover diamond resources in the United States has improved. However, evaluation of deposits already discovered will take several more years. Technology has been developed to synthesize diamond for industrial use worldwide in the range of sizes of crushing bort, powder, and dust. World resources of natural industrial diamond in the stone-size range are unknown.

11. **Substitutes:** Competitive materials were cubic boron nitride, fused aluminum oxide, and silicon carbide (as manufactured abrasive materials) and garnet, emery, and corundum (as natural abrasive minerals). Synthesized polycrystalline diamond was competitive with natural stones in many applications. Research continued on additional uses of synthetic polycrystalline compacts and shapes as substitutes for stones.

Table A-21.—*Continued*

MANGANESE

(Data in thousand short tons, gross weight, unless noted)

1. **Domestic production and use:** Manganese ore containing 35% or more manganese was not produced domestically in 1988. Manganese ore was consumed mainly by about 20 firms with plants principally in the Eastern and Midwestern United States. The largest portion of ore consumption was related to steel production directly through the making of pig iron and indirectly through upgrading of ore to ferroalloys and metal. Most of the remaining ore was used for such nonmetallurgical purposes as producing dry cell batteries. Overall end uses of manganese were construction, 24%; transportation, 15%; machinery, 11%; and other, 50%.

2. **Salient statistics—United States:**[1]

	1984	1985	1986	1987	1988[e]
Production, mine[2]	—	—	—	—	—
Imports for consumption:					
Manganese ore	338	387	463	340	535
Ferromanganese	409	367	396	368	500
Exports:					
Manganese ore	238	56	42	63	85
Ferromanganese	7	7	4	3	4
Shipments from Govt. stockpile excesses:[3]					
Manganese ore	65	191	150	228	90
Ferromanganese	(24)	—	(81)	—	(52)
Consumption, reported:[4]					
Manganese ore	615	[e]545	[e]500	[e]521	600
Ferromanganese	492	466	376	409	480
Consumption, apparent, manganese[5]	627	698	730	692	825
Price, average value, 46%–48% Mn metallurgical ore, dollars per ltu cont. Mn, c.i.f. US ports	1.42	1.43	1.34	1.29	1.80
Stocks, producer and consumer, yearend:					
Manganese ore	582	[e]589	[e]455	[e]470	425
Ferromanganese	159	100	93	48	40
Net import reliance[6] as a percent of apparent consumption	98	100	100	100	100

3. **Recycling:** Scrap recovery specifically for manganese was insignificant. However, considerable manganese was recycled through processing operations as a minor component of ferrous and nonferrous scrap and steel slag.

4. **Import sources (1984–87):** [7]Manganese ore: Gabon, 44%; Brazil, 23%; Australia, 19%; other, 14%. Ferromanganese: Republic of South Africa, 36%; France, 29%; other, 35%. Manganese contained in all manganese imports: Republic of South Africa, 27%; France, 14%; Gabon, 14%; Brazil, 11%; other 34%.

5. **Tariff:**

Item	Number	Most favored nation (MFN) 1/1/89	Non-MFN 1/1/89
Ore and concentrate	2602.00.00	Free	2.2¢/kg of contained Mn
High-carbon ferromanganese	7202.11.50	1.5% ad val.	10.5% ad val.
Metal	8111.00.45	14% ad val.	20% ad val.

6. **Depletion allowance:** 22% (Domestic); 14% (Foreign).

---

[e]. Estimated.
[1] Manganese content typically ranges from 35% to 54% for manganese ore and from 74% to 95% for ferromanganese.
[2] Excludes manganiferous ore containing less than 35% manganese, which accounts for about 2% or less of apparent consumption of manganese.
[3] Net quantity including effect of stockpile upgrading program. Data in parentheses denote stockpile acquisitions.
[4] Total manganese consumption cannot be approximated from consumption of manganese ore and ferromanganese because of the use of ore in making manganese ferroalloys and metal.
[5] Thousand short tons, manganese content. Based on estimates of average content for all significant components except imports for which content is reported.
[6] Defined as imports – exports + adjustments for Government and industry stock changes.
[7] Thousand short tons, manganese content; see page 186 for definitions.

Table A-21. MANGANESE—*Continued*

7. **Government stockpile:** Data tabulated below pertain to uncommitted inventories of stockpile-grade material. The stockpile contained further small quantities of natural battery ore, chemical ore, and metallurgical ore, that had been sold but not yet shipped. For these kinds of ore, the stockpile also contained about 34,000 tons, 90 tons, and 919,000 tons, respectively, of nonstockpile-grade material. At midyear, the General Services Administration exercised its option under the existing contract to extend upgrading of metallurgical manganese ore in the stockpile into high-carbon ferromanganese through 1989. In 1989, 103,000 tons of ore was to be converted into 58,600 tons of high-carbon ferromanganese.

Stockpile Status—9-30-88

Material	*Goal*	*Total inventory*	*Excess to goal authorized for disposal*	*Sales, 9 months*
Battery: Natural ore	62	169	11	2
Synthetic dioxide	25	3	—	—
Chemical ore	170	172	—	—
Metallurgical ore	2,700	2,026	—	—
Ferromanganese:				
High-carbon	439	757	—	—
Medium-carbon	—	29	—	—
Silicomanganese	—	24	—	—
Electrolytic metal	—	14	—	—

8. **Events, trends, and issues:** US prices rose significantly for all main forms of manganese used metallurgically, most notably for high-carbon ferromanganese and ore. Ore price and foreign trade, the latter as measured by manganese content of US imports and exports, were at the highest levels in nearly a decade, based on a partial year's data. Means for easing ore exporting were being implemented in Gabon and the Republic of South Africa.

It is estimated that in 1989 domestic mine production of manganese ore[2] will be zero and that US apparent consumption of manganese will be 760,000 tons, content basis. The majority of total US demand will be supplied by manganese imports in both ore and upgraded forms.

Manganese is an essential element for man and animals that can be harmful when in excess. Thus, manganese can be an industrial poison, but is not a hazard generally. The Environmental Protection Agency (EPA) found as of August 1985 that ambient air concentrations of manganese posed no significant risk to public health. In 1988, EPA requested data for certain manganese materials to make a preliminary ranking and screening of substances identified as potential air pollutants. EPA included manganese metal and chemical compounds among a number of substances whose release to the environment must be reported annually.

9. **World mine production, reserves, and reserve base:**

	Mine production		**Reserves[7]**	**Reserve base[7]**
	1987	**1988[e]**		
United States	—	—	—	—
Australia	2,043	2,000	75,000	168,000
Brazil	[e]2,650	2,800	20,900	69,000
Gabon	[e]2,650	2,650	110,000	190,000
India	1,436	1,450	20,000	30,000
Mexico	425	470	3,700	8,600
South Africa, Republic of	3,188	3,500	407,000	2,900,000
Other Market Economy Countries	[e]465	495	Small	Small
China	[e]1,760	1,760	15,000	32,000
USSR	[e]10,300	10,300	325,000	500,000
Other Centrally Planned Economies	[e]187	185	Small	Small
World Total (rounded)	[e]25,100	25,600	1,000,000	3,900,000

10. **World resources:** Identified land-based resources are large but are irregularly distributed. The USSR and the Republic of South Africa account for more than 80% of the world's identified resources; the Republic of South Africa accounts for more than 75% of the market economy countries. Extensive marine accumulations of manganese as oxide nodules on ocean floors and as oxide crusts at shallower depths may have future commercial significance.

11. **Substitutes:** There is no satisfactory substitute for manganese in its major applications.

Table A-21.—*Continued*

PLATINUM-GROUP METALS

(Platinum, palladium, rhodium, ruthenium, iridium, osmium)
(Data in thousand troy ounces, unless noted)

1. **Domestic production and use:** Concentrates containing platinum-group metals (PGM) were exported to Belgium for smelting and refining. In addition, refined PGM were recovered as byproducts of copper refining by one company in Texas. Secondary metal was refined by about 24 firms, mostly on the east and west coasts. PGM were sold by at least 90 processors and retailers, largely in the Northeast, and were distributed among using industries as follows: automotive, 45%; electrical and electronic, 22%; dental and medical, 12%; chemical, 6%; other, 15%. The automotive, chemical, and petroleum-refining industries used PGM mainly as catalysts. The other industries used PGM in a variety of ways that took advantage of their chemical inertness and refractory properties.

2. **Salient statistics—United States:**

	1984	1985	1986	1987	1988[e]
Production: Mine	15	W	W	W	W
Primary refined	24	7	4	6	5
Secondary					
Nontoll-refined	340	259	354	165	200
Toll-refined	1,157	1,038	1,155	1,444	1,500
Imports for consumption:					
Refined	3,928	3,438	3,727	3,179	3,300
Total	4,474	3,990	4,477	3,807	3,900
Exports: Refined	599	526	382	432	800
Total	1,162	889	751	708	1,000
Consumption:					
Reported, sales to industry	2,200	2,271	2,080	1,944	2,200
Apparent[1]	3,299	3,358	3,536	2,969	2,700
Price, dealer, average, dollars per ounce:					
Platinum	357	291	461	553	500
Palladium	148	107	116	130	120
Stocks, industry, yearend	1,319	1,129	1,292	1,235	1,200
Shipments from Govt. stockpile excesses[2]	(9)	(2)	—	—	—
Employment, refinery[e]	400	400	400	400	400
Net import reliance[3] as a percent of apparent consumption	89	92	90	94	93

3. **Recycling:** About 200,000 ounces of PGM were refined from scrap on a nontoll basis. The quantity of toll-refined secondary was much larger, amounting to 1.5 million ounces. Several large companies collected used catalytic converters from automobile salvage yards and muffler shops, recycled the outer shell of the converters, and shipped the catalyst and substrate to US and Japanese companies for extraction of platinum, palladium, and rhodium metals.

4. **Import sources (1984–87):** Republic of South Africa, 44%; United Kingdom, 16%; USSR, 9%; other, 31%.

5. **Tariff:**

Item	Number	Most favored nation (MFN) 1/1/89	Non-MFN 1/1/89
Unwrought, semimanufactured, or powdered forms	7110	Free	Free
Waste and scrap	7112.20	Free	Free

6. **Depletion allowance:** 22% (Domestic); 14% (Foreign).

---

[e] Estimated. W Withheld to avoid disclosing company proprietary data.
[1] Domestic mine production plus nontoll secondary production plus net import reliance (see footnote 3). Mine production for 1985–88 excluded from the calculation in order to avoid disclosing company proprietary data.
[2] Data in parentheses denote stockpile acquisitions.
[3] Defined as refined imports – refined exports + adjustments for Government and industry stock changes.
[4] See page 186 for definitions.
[5] Excludes US production.

Table A-21. PLATINUM-GROUP METALS—*Continued*

7. **Government stockpile:** No PGM were purchased in 1988.

Stockpile Status—9-30-88

Material	*Goal*	*Total inventory*	*Excess to goal authorized for disposal*	*Sales, 9 months*	*Purchases, 9 months*
Platinum	1,130	440	—	—	—
Palladium	3,000	1,262	—	—	—
Iridium	98	30	—	—	—

In addition to quantities shown above, the stockpile contains 13,043 ounces of nonstockpile-grade platinum and 2,214 ounces of nonstockpile-grade palladium.

8. **Events, trends, and issues:** The Bureau of Mines published an Open File Report (OFR 19-88) entitled "Estimated Direct Economic Impacts of a US Import Embargo on Strategic and Critical Minerals Produced in South Africa." In the report, the Bureau estimated the average annual cost of an embargo of South African platinum, palladium, and rhodium to be $1.7 billion. One of the conclusions of the report was that world sources for platinum and rhodium, other than South Africa, probably would not meet US industrial demand. The Bureau published another Open File Report (OFR 54-88) entitled "Estimated Impacts on US Gross National Product (GNP) and Employment Resulting from a US Embargo on South African Platinum-Group Metal Supplies." The Bureau concluded that an embargo on South African rhodium supplies would reduce automobile production and employment and result in estimated GNP losses averaging $12 billion annually from 1988–92.

   Two new platinum coins, the Australian Koala and the Canadian Maple Leaf, were marketed in North America. Both were 99.95%-pure platinum and considered legal tender. They were available in weights of 1 ounce, 1/2 ounce, 1/4 ounce, and 1/10 ounce. The prices of the coins were based on the current market price of platinum plus a small premium to cover minting and handling costs. In 1988, investment demand represented more than 15% of world demand for platinum and was the third largest end use for platinum after automobile catalysts and jewelry.

9. **World mine production, reserves, and reserve base:**

	Mine production		**Reserves[4]**	**Reserve Base[4]**
	1987	**1988[e]**		
United States	W	W	8,000	25,000
Canada	434	400	8,000	9,000
South Africa, Republic of	4,220	4,500	1,600,000	1,900,000
Other Market Economy Countries	117	100	1,000	1,000
USSR	3,900	3,900	190,000	200,000
World Total (rounded)	[5]8,700	[5]8,900	1,810,000	2,140,000

10. **World resources:** World resources of PGM are estimated to be 3.3 billion ounces. US resources are estimated at 300 million ounces.

11. **Substitutes:** The feasibility of substituting other materials for PGM is greatest for electronic applications and least for catalyst applications. Tin-lead alloys can substitute for palladium-silver alloys used in multilayer capacitors. In dental crowns, gold, silver, and sometimes titanium can substitute for palladium alloys. Research continues into the use of base metals clad or plated with PGM and into other ways of minimizing the quantities of PGM used in various industrial applications. A company in Japan is substituting palladium alloyed with lanthanum for platinum and rhodium in three-way automobile catalysts.

Table A-21.—*Continued*

TANTALUM

(Data in thousand pounds of tantalum content, unless noted)

1. **Domestic production and use:** There has been no significant tantalum-mining industry since 1959. Most metal, alloys, and compounds were produced by five companies with six plants; tantalum units were obtained from imported concentrates and tin slags, and from both foreign and domestic scrap. Total estimated value of domestic shipments of metal, alloys, and compounds was about $145 million. Consumption in the form of metal powder, ingot, fabricated forms, and compounds and alloys had the following end uses: electronic components, 60%; transportation, 15%; machinery, 11%; and other, 14%.

2. **Salient statistics—United States:**

	1984	1985	1986	1987	1988[e]
Production, mine	—	—	—	—	—
Imports for consumption, concentrate, tin slags, and other[1]	W	W	W	W	W
Exports, concentrate, metal, alloys, waste, and scrap[e]	383	320	312	376	330
Shipments from Govt. stockpile excesses	—	[2](254)	—	—	—
Consumption: Reported, raw material[e]	1,300	1,100	W	W	NA
Apparent	1,680	800	820	840	910
Price, tantalite, dollars per pound[3]	30.66	27.58	19.44	22.18	39.00
Stocks, industry, processor, yearend	W	W	W	W	W
Employment, processor[e]	500	500	450	450	450
Net import reliance[4] as a percent of apparent consumption	92	89	91	85	89

3. **Recycling:** Combined prompt industrial and obsolete scrap consumed was more than 10% of raw materials consumed. Generation of new scrap was estimated to be around 300,000 pounds, most of which was consumed by processors.

4. **Import sources (1984–87):** Thailand, 32%; Brazil, 11%; Australia, 8%; Canada, 4%; other, 45%.

5. **Tariff:**

Item	Number	Most favored nation (MFN) 1/1/89	Non-MFN 1/1/89
Synthetic tantalum-columbium concentrate	2615.90.3000	Free	30.0% ad val.
Tantalum concentrate	2615.90.6060	Free	Free
Potassium fluotantalate	2826.90.0000	3.1% ad val.	25.0% ad val.
Tantalum unwrought:			
Waste and scrap	8103.10.3000	Free	Free
Metal	8103.10.6030	3.7% ad val.	25.0% ad val.
Alloys	8103.10.6090	4.9% ad val.	25.0% ad val.
Tantalum, wrought	8103.90.0000	5.5% ad val.	45.0% ad val.

6. **Depletion allowance:** 22% (Domestic); 14% (Foreign).

[e] Estimated. NA Not available. W Withheld to avoid disclosing company proprietary data.
[1] Metal, alloys, and synthetic concentrates; exclusive of waste and scrap.
[2] Data in parentheses denote stockpile acquisitions.
[3] Average value, contained tantalum pentoxides, 60% basis.
[4] Defined as imports – exports + adjustments for Government and industry stock changes.
[5] Excludes production of tantalum contained in tin slags reported by the Tantalum-Niobium International Study Center (TIC) as 543,000 pounds in 1987; data not available for 1988.
[6] See page 186 for definitions.
[7] Exclude centrally planned economies.

Table A-21. TANTALUM—*Continued*

7. **Government stockpile:** In February 1988, the President issued an Executive order designating the Secretary of Defense as the manager of the National Defense Stockpile. In July 1988, the Defense Logistics Agency of the Department of Defense officially assumed all stockpile management duties, taking over that role from the Federal Emergency Management Agency. In addition to quantities shown below, the stockpile contains a negligible quantity in nonstockpile-grade metal and 1,152,000 pounds in nonstockpile-grade minerals.

Stockpile Status—9-30-88

Material	*Goal*	*Total inventory*	*Excess to goal authorized for disposal*	*Sales, 9 months*
Carbide powder	—	29	—	—
Metal	—	201	—	—
Minerals	8,400	1,686	—	—

8. **Events, trends, and issues:** Demand for tantalum continued to improve in 1988, aided by the strong consumption of capacitor powder in the electronics sector. Imports for consumption of tantalum feed material were up substantially from 1987 to the highest level since 1984. However, the price escalation for tantalum feed materials was a major concern, because the quoted price for tantalite nearly doubled during 1988. The published spot price for tantalum mineral concentrates, which began the year at $26 per pound of contained pentoxide, was up to $37 by midyear and was being quoted in the fourth quarter at around $46, the highest level since third quarter 1981. Depending on the size of the order-contract, industry sources indicated that tantalum mill products and powders sold in the range of $100 to $160 per pound. It is estimated that in 1989 domestic mine production of tantalum will be zero and that US apparent consumption will be less than 1 million pounds.

   Tantalum supply received a boost in September 1988 with the startup of tantalite production by the sole Canadian producer. This mine operation had been suspended since yearend 1982. Plans were also announced to rebuild the Thai columbium- and tantalum-processing plant that was destroyed by fire on June 23, 1986. The fire occurred just prior to commissioning of the plant for operation.

   There are no known uncontrollable health hazards connected with production or fabrication of tantalum metals and compounds. Fumes, gases, dust, and low-level radiation generated by tantalum extraction plants can be controlled by modern technology

9. **World mine production, reserves, and reserve base:**

	Mine production[e5]		**Reserves**	**Reserve base[6]**
	1987	**1988**		
United States	—	—	—	Negligible
Australia	115	120	10,000	20,000
Brazil	190	200	2,000	3,000
Canada	—	60	4,000	5,000
Malaysia	33	40	2,000	4,000
Nigeria	2	10	7,000	10,000
Thailand	107	110	16,000	20,000
Zaire	31	30	4,000	10,000
Other Market Economy Countries	45	60	3,000	4,000
Centrally Planned Economies	NA	NA	NA	NA
World Total[7]	523	630	48,000	76,000

10. **World resources:** Most of the world's resources of tantalum occur outside the United States. On a worldwide basis, identified resources of tantalum are considered adequate to meet projected needs. These resources are largely in Australia, Brazil, Canada, Egypt, Malaysia, Nigeria, Thailand, and Zaire. The United States has about 3 million pounds of tantalum resources in identified deposits, which were considered uneconomic at 1988 prices.

11. **Substitutes:** The following materials can be substituted for tantalum, but usually with less effectiveness: columbium in superalloys and carbides; aluminum and ceramics in electronic capacitors; glass, titanium, zirconium, columbium, and platinum in corrosion-resistant equipment, and tungsten, rhenium, molybdenum, iridium, hafnium, and columbium in high-temperature applications.

Table A-21.—*Continued*

TUNGSTEN

(Data in metric tons of tungsten content, unless noted)

1. **Domestic production and use:** In 1988, production of tungsten concentrate occurred at only one mine. The mine and mill in California was operated at a small fraction of its capacity, providing supplemental feed material to the company's nearby ammonium paratungstate plant. A second mine in California was prepared for reopening. Production of concentrate at this site was delayed while management awaited final approval from the Securities and Exchange Commission to issue common stock to support the operation of the facilities. End uses of tungsten were metalworking, mining, and construction machinery and equipment, 67%; lamps and lighting, 12%; transportation, 7%; electrical machinery and equipment, 7%; and other, 7%. The total estimated value of tungsten consumed in 1988 was $200 million.

2. **Salient statistics—United States:**

	1984	1985	1986	1987	1988[e]
Production, mine shipments	1,173	983	817	34	230
Imports for consumption, concentrate	5,807	4,746	2,522	4,414	7,200
Exports, concentrate	129	124	34	2	145
Shipments from Govt. stockpile excesses, concentrate	1,368	902	301	708	524
Consumption: Reported concentrate	8,577	6,838	4,804	5,506	7,750
Apparent concentrate, scrap, metal	10,164	8,211	7,767	7,951	10,044
Price, concentrate, dollars/mtu WO_3,[1] average:					
US market	85	68	46	50	60
European market	87	71	53	53	58
Stocks, producer and consumer, yearend, concentrate	1,005	1,137	523	350	530
Employment, mine and mill	287	275	187	77	125
Net import reliance[2] as a percent of apparent consumption	70	68	70	79	75

3. **Recycling:** During 1988, the quantity of purchased scrap represented about 2,300 tons.

4. **Import sources (1984–87):** China, 26%; Canada, 11%; Bolivia, 11%; Federal Republic of Germany, 7%; other, 45%.

5. **Tariff:**

Item	Number	Most favored nation (MFN) 1/1/89	Non-MFN 1/1/89
Ore and concentrate	2611.0000	17¢/lb W cont.[3]	50¢/lb W cont.
Ferrotungsten	7202.8000	5.6% ad val.	35.0% ad val.
Unwrought lumps, grains, and powders	8101.1000	10.5% ad val.	58.0% ad val.
Ammonium tungstate	2841.8000	10.0% ad val.	49.5% ad val.
Tungsten carbide	2849.9030	10.5% ad val.	55.5% ad val.

6. **Depletion allowance:** 22% (Domestic), 14% (Foreign).

7. **Government stockpile:** Disposal of nonstockpile-grade ore and concentrate for the first 9 months in support of the ferroalloy upgrading program totals 524 tons. Authority for disposal of an additional 454 tons of ore and concentrate for this program was in effect as of October 1. There were no sales of nonstockpile-grade concentrate acquired under the Defense Production Act (DPA) of 1950. About 5 tons of DPA material remain available for disposal. In addition to the stockpile-grade quantities shown in this section, the stockpile contains the following nonstockpile-grade materials (in tons): ore and concentrate, 10,526; metal powder, 151; ferrotungsten, 537; and carbide powder, 51.

Stockpile Status—9-30-88

Material	*Goal*	*Total inventory*	*Excess to goal authorized for disposal*	*Sales, 9 months*
Ore and concentrate	25,152	24,573	72	—
Metal powder	726	711	—	—
Ferrotungsten	—	381	—	—
Carbide powder	907	871	—	—

[e] Estimated.

[1] A metric ton unit (mtu) of tungsten trioxide (WO_3) contains 7.93 kilograms of tungsten.

[2] Defined as imports – exports + adjustments for Government and industry stock changes.

[3] Waived Oct. 1, 1988 through Dec. 31, 1990 per Omnibus Trade Bill signed Aug. 23, 1988.

[4] See page 186 for definitions.

Table A-21. TUNGSTEN—*Continued*

8. **Events, trends, and issues:** In 1988, tungsten mines in most market economy countries continued to be closed or continued to be operated below capacity. Prices for tungsten concentrates rose an average of 15% above that of 1987, since the demand for tungsten materials increased. Reported US consumption of tungsten concentrate increased an estimated 40% largely owing to the strength of the cemented carbide tool insert market. Nearly all concentrates consumed in the United States were supplied from imports and shipments from US Government Defense Stockpile excesses. Net import reliance, as a percent of apparent consumption of tungsten material, remained high, although it declined by 4% in 1988 to a level of 75%.

   According to provisions in the orderly marketing agreement signed with China, the Office of the US Trade Representative adjusted downward the import quotas for ammonium paratungstate (APT) and tungstic acid. The adjustment for 1988 was about 950,000 lb., partially compensating for the excessive quantity of imports that arrived during the 1987 base period. APT production in the United States increased by an estimated 37% in 1988 compared with that of 1987.

   In June 1988, the European Commission received formal complaints from the European Economic Community's (EEC) tungsten-industry members regarding excessive imports of tungsten products from China. Separate dossiers covering both ores and concentrates and upgraded products were submitted by the EEC.

   A new trade association, the International Tungsten Industry Association (ITIA), was formally inaugurated in February 1988. It will serve the interests of the tungsten industry as a whole. Objectives included compiling and disseminating international tungsten statistics, promoting new uses for tungsten, monitoring environmental and health issues, and organizing symposia and seminars.

9. **World mine production, reserves, and reserve base:**

	Mine production		Reserves[4]	Reserve base[4]
	1987	1988[e]		
United States	34	230	150,000	210,000
Australia	1,150	1,200	130,000	150,000
Austria	1,250	1,500	15,000	20,000
Bolivia	500	1,200	45,000	110,000
Brazil	672	700	20,000	20,000
Burma	425	500	15,000	34,000
Canada	—	—	350,000	493,000
France	—	—	20,000	20,000
Korea, Republic of	2,500	2,000	58,000	77,000
Portugal	1,500	1,700	26,000	26,000
Thailand	660	600	30,000	30,000
Other Market Economy Countries	1,796	1,800	232,000	290,00
China	18,000	18,000	1,200,000	1,560,000
USSR	9,200	9,200	280,000	400,000
Other Centrally Planned Economies	2,545	2,500	85,000	105,000
World Total	40,232	41,130	2,656,000	3,545,000

10. **World resources:** More than 90% of the world's estimated tungsten resources are located outside the United States, with about 50% located in China. Besides China and the United States, other areas with significant resource potential are in Australia, Austria, Bolivia, Brazil, Burma, Canada, North Korea, Peru, Portugal, the Republic of Korea, Spain, Thailand, Turkey, and the USSR.

11. **Substitutes:** Advancements in carbide- and oxide-coatings technology have improved the cutting and wear resistance of cemented tungsten carbide tool inserts. Coatings are estimated to be used on 30% to 35% of the inserts. The extended wear capability of the inserts decreases the replacement rate and, hence, the growth of tungsten consumption. Gradual increases in the substitution of cemented tungsten carbide base products by titanium carbide base cutting tools, by ceramic cutting tools and wear parts, and by polycrystalline diamond have also occurred.

Table A-21.—*Continued*

VANADIUM

(Data in thousand pounds of vanadium content, unless noted)

1. **Domestic production and use:** The US vanadium industry consisted of 10 firms in 1988, but only 6 had active extraction operations. Raw materials included Idaho ferrophosphorous slag, petroleum residues, utility ash, and imported iron slags. The chief use of vanadium was as an alloying agent for iron and steel. It was also important in the production of aerospace titanium alloys, and as a catalyst for production of sulfuric acid. About 140 plants throughout the United States reported consumption in 1988. Major end-use distribution was as follows: machinery and tools, 36%; transportation, 27%; building and heavy construction, 25%; and other, 12%.

2. **Salient statistics—United States:**

	1984	1985	1986	1987	1988[e]
Production:					
Mine, recoverable basis	3,234	W	W	W	W
Mill, recovered basis[1]	5,240	W	W	W	W
Petroleum residues, recovered basis	3,401	5,390	4,660	5,106	5,217
Imports for consumption:					
Ores, slag, residues	1,266	605	4,026	4,528	4,841
Vanadium pentoxide, anhydride	297	22	824	457	380
Ferrovanadium	2,341	1,557	1,189	684	670
Exports:					
Ores	24	5	172	—	—
Vanadium pentoxide, anhydride	4,158	1,710	1,730	1,636	1,660
Other compounds	366	386	346	575	587
Ferrovanadium	656	635	820	685	695
Shipments from Govt. stockpile[2]	(136)	(224)	—	—	—
Consumption: Reported	9,522	9,766	8,616	9,306	9,678
Apparent[3]	11,458	W	W	W	W
Price, average, dollars per pound V_2O_5	3.50	3.50	3.50	3.50	4.35
Stocks, producer and consumer, yearend	7,540	6,418	5,463	4,535	4,480
Employment, mine and mill	660	580	585	699	730
Net import reliance[4] as a percent of apparent consumption	54	W	W	W	W

3. **Recycling:** Some tool steel scrap was recycled primarily for its vanadium content. Vanadium was also recycled as a minor component of scrap iron and steel alloys, which were used principally for their iron content.

4. **Import sources (1984–1987):**[5] Republic of South Africa, 36%; South America, 26%; European Communities (EC), 11%; Canada, 9%; Austria, 5%; other, 13%.

5. **Tariff:**

Item	Number	Most favored nation (MFN) 1/1/89	Non-MFN 1/1/89
Ores and concentrates	2615.90.60.90	Free	Free
Slag	2619.00.90.30	Free	Free
Ash and residues	2620.50.00.00	Free	Free
Vanadium pentoxide anhydride	2825.30.00.10	16.0% ad val.	40% ad val.
Vanadates	2841.90.10.00	11.2% ad val.	40% ad val.
Ferrovanadium	7202.92.00.00	4.2% ad val.	25% ad val.
Aluminum vanadium master alloys	7601.20.90.30	Free	10.5% ad val.
Waste and scrap	8112.40.30.00	Free	Free
Vanadium oxides and hydroxides, other	2825.30.00.50	16.0% ad val.	40% ad val.

[e]. Estimated. W Withheld to avoid disclosing company proprietary data.
[1] Produced from domestic materials.
[2] Data in parentheses denote stockpile acquisitions.
[3] Includes processing losses from low-grade imports.
[4] Defined as imports – exports + adjustments for Government and industry stock changes.
[5] The EC, Canada, and Austria produced vanadium alloys and chemicals solely from imported raw materials.
[6] Excludes US production.
[7] See page 186 for definitions.
[8] Includes data for Bophuthatswana.

Table A-21. VANADIUM—*Continued*

6. **Depletion allowance:** 22% (Domestic), 14% (Foreign).
7. **Government stockpile:**

Stockpile Status—9-30-88

Material	*Goal*	*Total inventory*	*Excess to goal authorized for disposal*	*Sales, 9 months*	*Purchases, 9 months*
Vanadium pentoxide	15,400	1,442	—	—	—
Ferro-vanadium	2,000	—	—	—	—

8. **Events, trends, and issues:** Amid continued tight supply and strong demand for vanadium products, the price of vanadium pentoxide increased at the rate of about 4% per month through the first 9 months of 1988. Increased demand for vanadium products came mostly from the steel industry which had the highest first half shipments since 1980. Although all major steel consuming sectors increased purchases, the oil and gas industry reported the biggest change, up more than 120% from the same period in 1987. The extraordinary demand for vanadium created shortages of vanadium raw materials, and in turn, raised fears about the impact of any further sanctions against South Africa, a major producer of vanadium.

   The Supreme Court ruled that the Government does not have to place restrictions on imported uranium to protect domestic producers. The unanimous decision overturned a decision banning Federal officials from enriching foreign uranium, particularly from Canada and Australia.

   The Nuclear Regulatory Commission (NRC) began reviewing reclamation plans for two uranium-vanadium mills, which are scheduled to be decommissioned. The reclamation will take about 7 years to complete at a cost of about $6 million.

9. **World mine production, reserves, and reserve base:**

	Mine production[6]		**Reserves[7]**	**Reserve base[7]**
	1987	**1988[e]**		
United States	W	W	300,000	4,800,000
Australia	—	—	70,000	540,000
Finland	—	—	—	200,000
South Africa, Republic of[8]	37,400	38,500	1,900,000	17,200,000
Other Market Economy Countries	—	—	—	1,200,000
China	10,000	9,000	1,340,000	3,600,000
USSR	21,200	19,500	5,800,000	9,000,000
World Total (may be rounded)	68,600	67,000	9,410,000	36,500,000

10. **World resources:** World resources of vanadium exceeded 140 billion pounds. Vanadium occurs in deposits of titaniferous magnetite, phosphate rock, and uraniferous sandstones and siltstones where it constitutes less than 2% of the host rock. Significant amounts are also present in bauxite and carboniferous materials such as crude oil, coal, oil shale, and tar sands. Because vanadium is generally recovered as a byproduct or coproduct, demonstrated world resources of the element are not fully indicative of available supplies. While domestic resources are adequate to supply current domestic needs, a substantial part of US demand is currently met by foreign material because of price advantages.
11. **Substitutes:** Steels containing various combinations of other alloying elements can be substituted for steels containing vanadium. Among the various metals that are interchangeable with vanadium are columbium, molybdenum, manganese, titanium, and tungsten. There is currently no acceptable substitute for vanadium in aerospace titanium alloys. Platinum and nickel can replace vanadium compounds as a catalyst in some chemical processes.

Table A-22. Technical Prospects for Reducing US Import Vulnerability for Chromium, Cobalt, Manganese, and the Platinum Group Metals

	Chromium	*Cobalt*	*Manganese*	*Platinum group*
US apparent consumption (1982)[a]	319,000 short tons	5,592 short tons (11,184,000 lbs.)	672,000 short tons	61 short tons (1,787,000 troy ounces)
U.S. import dependence (1982)[a]	85%	92%	99%	80%
Recycling in 1982 (purchased scrap)[b]	12%	8%	Not estimated[b]	19% (excludes toll refined)
Domestic production in 1982[a]	0%	0%	About 2% of apparent consumption	0.4% of apparent consumption
Price in 1982[a]	Turkish chromite: $100/short ton ($110/metric) South African: $47/short ton ($52/metric).	Cathode 99 % Co: $12.90/pound	About $1.58–$1.68 per long ton unit (22.4 pounds) of 46 to 48% of Mn metallurgical ore.	Dealer price: platinum—$327/troy oz.; palladium—$67/troy oz.

Office of Technology Assessment, 1984.

[a] US Department of Interior, Bureau of Mines, *Mineral Commodity Summaries 1984*, (Washington, DC: US Government Printing Office, 1983). Figures on apparent consumption may differ from apparent consumption figures in table A or ch. 1 because of different reporting conventions used in different Bureau of Mines statistical series.

[b] US Department of Interior, Bureau of Mines, *Mineral Commodity Summaries 1983*, (Washington, DC: US Government Printing Office, 1982). Estimates of purchased scrap as a percentage of apparent consumption in 1982 are preliminary estimates which may be subject to revision. Appreciable quantities of manganese contained in ferrous and nonferrous scrap and steelmaking slag are recycled. However, scrap recovery specifically to recycle manganese is minimal.

[c] US Department of Interior, Bureau of Mines, *Minerals Yearbook: Vol. I, Metals and Minerals 1982* (Washington, DC: US Government Printing Office, 1983), table 10, pp. 54–55.

[d] US Department of Interior, Bureau of Mines, "Mineral Commodity Profiles 1983" (chromium, cobalt, manganese, and platinum group metals).

Table A-22—*Continued*

	Chromium	*Cobalt*	*Manganese*	*Platinum group*
Value of imports into the United States in 1982[c] (including gross weight)	$120 million	$143 million	$197 million	$554 million
Import sources (1979–1982)[a] (Bold-faced countries are primary producers)	Chromite: South Africa (48%); USSR (17%); Philippines (13%); Other (22%); Ferrochromium: South Africa (44%); Yugoslavia (9%); Zimbabwe (9%); Other (38%).	Zaire (37%); Zambia (13%); Canada (8%); Belgium-Luxembourg (8%); Japan (7%); Norway (7%); Other (13%).	Ore: South Africa (33%); Gabon (25%); Australia (20%); Brazil (12%); Other (9%). Ferromanganese: South Africa (43%); France (26%); Other (31%).	South Africa (56%); USSR (18%); UK (11%); Other (17%).
Location of major world reserves[d]	South Africa and Zimbabwe (92%); the remaining 8%, distributed among more than 10 countries, represents reserves in excess of 100 million short tons of chromite. Chromite typically ranges from 22 to 38% chromium content.	Zaire (50%); Zambia (13%); Cuba (7%); USSR (5%); the remaining 25% of world reserves (2 million short tons) is distributed among 12 countries.	South Africa (40%); USSR (37%); remaining 23% (distributed among 8 countries) represents reserves in excess of 200 million short tons.	South Africa (79%); USSR (19%); remaining 2% (20 million troy ounces) is in Canada, the US, and Colombia.

Table A-22—*Continued*

	Chromium	*Cobalt*	*Manganese*	*Platinum group*
Prospects for increased substitution	Very good for noncritical applications; for critical applications, extensive applied R&D will be needed. Basic research breakthroughs may be needed to develop substitutes in the most critical applications.	Good for many applications; additional R&D needed for critical applications such as superalloys. Qualification requirements may limit practical use of these substitutes.	Poor—substitutes for Mn in steelmaking (which accounts for 90% of consumption) are not promising.	Alternatives to PGM exist in many applications; in critical applications, prospects for direct substitution are poor in the near and medium term.
Potential for displacement by advanced materials	Fiber-reinforced plastics and some other composite materials may compete with stainless steels in some critical applications over time.	Reasonable prospects for incremental phase-in of advanced materials in near and medium term; long-term prospects (2010 and beyond) may be great in heat engine applications.	Limited: various composites may compete with steel in specialized applications.	Breakthrough in basic research is probably needed for other materials to replace PGM as a catalyst.

Table A-22—*Continued*

	Chromium	*Cobalt*	*Manganese*	*Platinum group*
Potential for increased recycling	Good—major obstacles are economic. Reduced downgrading, improved obsolete scrap recovery, and recovery of Cr values from steelmaking wastes appear to be the major opportunities, although data is out-of-date. Scrapped catalytic converter shells could become source of Cr; recovery of Cr from steelmaking wastes also may add to supplies.	Good—economic factors are primary impediments to recycling, although advanced technologies may have to be developed to maximize recovery of superalloy scrap. Obsolete superalloy scrap, spent catalysts, and other postconsumer uses will be a growing potential source of cobalt. For 1980, an estimated 2,350 short tons (4.7 million lbs) of Co was not recycled or was downgraded.	Poor—unless technical advances make Mn recovery from slag economical.	Excellent—recycling of PGM from automotive catalysts could provide 400,000 to 500,000 troy ounces of PGM annually in the mid-1990s. Obsolete electronic scrap could also be a major recycling source if high costs of disassembly are overcome.
Potential for more efficient use through design, processing, and manufacturing technologies.	Incremental gains, a major breakthrough akin to the AOD process, which reduced chromium losses in stainless steel production is not foreseen.	Very good—in superalloy production and parts fabrication; however, more efficient manufacturing will reduce scrap materials most preferred for recycling.	Good—manganese required per ton of steel could be reduced significantly by the year 2000 (from the current level of 36 lbs/ton to 25 lbs/ton).	Uncertain—will depend on basic research in catalyst chemistry.

Table A-22—*Continued*

	Chromium	*Cobalt*	*Manganese*	*Platinum group*
Prospects for production and processing:				
Potential for significant domestic production	Very poor—known deposits are low in quality and probably will not be exploitable except in a national emergency; however possibility of a new discovery of a promising deposit cannot be ruled out entirely. In addition, limited quantities of chromite may be produced as a by-product if nickel-cobalt laterites are developed.	Poor—without subsidy or major price rise. Good with subsidy. Maximum simultaneous development from existing mine sites could provide 10 million pounds (5,000 short tons) of cobalt annually for a 10- to 15-year period. Production economics varies by mine sites. Mine owners have cited prices ranging from $16 to $25 per pound as needed where cobalt is the primary mine product. At other sites, prices of nickel and copper need to be considered in determining whether coproduction would be profitable.	Very poor—known deposits very low in quality. Possible discovery of a promising deposit cannot be ruled out entirely.	Fair to good—initial plans to develop one domestic site anticipate palladium and platinum production equivalent to 9% of total US PGM consumption in 1982. A decision on the project is expected in 1985. Much higher production levels may be achievable if other sites are developed.

Table A-22—*Continued*

	Chromium	*Cobalt*	*Manganese*	*Platinum group*
Potential for supply diversification abroad	Poor to fair—probably would require US development assistance or incentives to promote production in Turkey, the Philippines, and several other countries.	Good prospects—but viability depends on copper and nickel markets because cobalt is a byproduct from mining of these materials. Improved technologies for laterite processing in the US may be transferable overseas.	Good—several alternative suppliers exist, including Mexico, Australia, Brazil, and Gabon.	Poor—unless major new discoveries are made (US has most promising unexploited deposit).
Potential to retain domestic processing capacity	Loss of processing capacity is expected to continue in near term with stabilization of domestic industry at a reduced level in long term to meet specialty steel industry needs.	Good—processing capacity is expanding due to recycling efforts.	Loss of processing capacity is expected to continue in near term with stabilization of industry at a reduced level to meet specialty steel industry needs.	Good—domestic refining capacity has increased to meet recycling needs.
Prospects for substantial reduction in current levels of US dependence by the year 2000 using the above alternatives	Fair	Good	Fair	Good

Table A-22—*Continued*

	Chromium	*Cobalt*	*Manganese*	*Platinum group*
Most promising technical routes to reducing US import dependence by 2000.	Substitution. Recycling Diversification of supplies	Recycling Design, processing, and manufacturing technologies Domestic production substitution	Continuation of conservation trends in steelmaking Diversification of supplies Replacement of steel in some applications by aluminum or plastics	Recycling of automobile catalysts and electronic scrap Domestic production

Table A-23. Potential US Cobalt Production

	Estimated Annual Production Capacity (Million Pounds of Recovered Cobalt)	*Estimated Minelife (Years)*	*Production Dependent on*
Blackbird Mine, Idaho	3.7	20	Cobalt price of about $16 per pound (1984)[a]
Madison Mine, Missouri	2	20	Cobalt price of about $25 (1981)[a]
Missouri Lead Belt, (tailings)	2	7	Cobalt price of about $20-25 (1984)[a]
Gasquet Mountain, California	2	18	Cobalt price of about $20 (1981)[a] plus Nickel $2 to $3 per pound Chromite $40 per ton
Duluth Gabbro, Minnesota	0.8–2	25	Copper, $1.50 per pound Nickel $4.00 (1975 data converted to January 1983 dollars)

[a]Year of estimate.

Blackbird—Noranda Mining, July 1984.
Madison—Anchutz Mining, July 1984.
Missouri Lead Belt—Amex estimates, July 1984.
Gasquet Mountain—California Nickel Corp., July 1984.
Duluth Gabbro—State of Minnesota, *Regional Copper-Nickel Study*, 1979.

Table A-24. Potential US Chromite Production

Known Resources by Deposit Type	Demonstrated Resources			Estimated Annual Production		
	Grade (Percent Cr_2O_3)	*Ore (Thousand Tonnes)*	*Chromium Content (Thousand Tonnes)*	*Ore (Thousand Tonnes)*	*Chromium Content (Thousand Tonnes)*	*Estimated Minelife (Years)*
Stratiform:						
Stillwater Complex:						
Mouat/Benboe	W	W	W	525	72[a]	46
Gish	15.0	500	51	175	16	3
Podiform:						
California:						
Bar Rick Mine	7.6	5,065	262	350	16	13
McGuffy Creek	W	W	W	788	NA	4
Pilliken Mine[b]	5.0	30,975	1,053	2,100	65	4

[a]Estimated assuming 15 percent grade.
[b]Inferred resources only.
W—information withheld for proprietary reasons.
NA—Data not available

Resources are ore grades, proposed mining rate, minelifes from US Department of the Interior, Bureau of Mines, *Chromium Availability—Domestic*, IC8895/1982. Gasquet Mountain data provided by California Nickel Corp.; balance calculated by OTA.

Table A-24—*Continued*

	Demonstrated Resources			*Estimated Annual Production*		
Known Resources by Deposit Type	*Grade (Percent Cr_2O_3)*	*Ore (Thousand Tonnes)*	*Chromium Content (Thousand Tonnes)*	*Ore (Thousand Tonnes)*	*Chromium Content (Thousand Tonnes)*	*Estimated Minelife (Years)*
Selad Creek/Emma Bell	5.0	4,546	155	563	17	9
Beach Sands:						
Southwest Oregon	5.6	10,827	412	1,000	35	11
Laterite	*Grade (Percent Chromium)*	*Proven Reserves (Thousand Tonnes)*		*Chromite Concentrates (Thousand Tonnes)*		
Gasquet Mountain	2.0	16,000	320	50	14	18

Table A-25. US Manganese Resources and Potential Production

Property Name by State	*Demonstrated Resource* Manganese Grade (Percent)	Ore (Thousand Tonnes)	Contained Manganese (Thousand Tonnes)	*Estimated Annual Mine Capacity* Ore (Thousand Tonnes)	Contained Manganese (Thousand Tonnes)	Estimated Minelife (Years)
Arizona:						
Hardshell Mine	15.0	5,896	804	536	73	11
Maggie Mine (Artillery Peak)	8.8	8,441	671	328	26	26
Colorado:						
Sunnyside Mine	10.0	24,909	2,264	635	58	39
Maine—Aroostock County:						
Maple Mtn/Hovey Mtn	8.9	260,000	20,965	4,263	344	61
North District	9.5	63,100	5,472	2,620	227	24
Minnesota:						
Cuyuna North Range (SW portion)	7.8	48,960	3,490	3,570	254	14

Resources, ore grade, proposed ore mining capacity from US Department of the Interior, Bureau of Mines, *Manganese Availability—Domestic*, IC8889/1982. Balance, calculated by OTA using that data.

Table A-25—*Continued*

Property Name by State	*Demonstrated Resource*			*Estimated Annual Mine Capacity*		
	Manganese Grade (Percent)	*Ore (Thousand Tonnes)*	*Contained Manganese (Thousand Tonnes)*	*Ore (Thousand Tonnes)*	*Contained Manganese (Thousand Tonnes)*	*Estimated Minelife (Years)*
Montana:						
Butte District (Emma Mine)	18.0	1,232	202	400	65	3
Nevada:						
Three Kids Mine	13.2	7,230	868	1,050	126	7
Total		419,768	34,737	13,401	1,174	

Table A-26. Potential US PGM Production

Resource/Mine	*Estimated Annual Production Capacity (Troy Ounces of Contained Metal)*		*Estimated Minelife (Years)*	*Production Dependent on*
Stillwater, Montana			10-25	Combined platinum-palladium price of about $220 per troy ounce (1984)[a]
Initial:	Palladium	136,000		
	Platinum	38,000		
	Total	175,000		
Additional expansion:	Palladium	340,000		
	Platinum	97,000		
	Total	437,000		
Goodnews Bay, Alaska			unknown	Platinum price of $600–700 per troy ounce (1984)
	Platinum	10,000		
Duluth Gabbro, Minnesota			25	Copper, $1.50 per pound
	Palladium	30,800– 92,400		Nickel, $4.00 (1975 data converted to January 1983 dollars)
	Platinum	6,800– 20,300		
	Total	37,600–112,700		

[a]Year of estimate.

Stillwater—Stillwater Mining, June 1984.
Goodnews Bay—Henson Properties, July 1984.
Duluth Gabbro—Calculated by OTA using preliminary results of Bureau of Mines research on Duluth ores; State of Minnesota, *Regional Copper-Nickel Study,* 1979, Silverman, et. al., OTA background study, 1983.

Appendix B
Ten Strategic Minerals

Andalusite

LARGE-PARTICLE, COARSE-GRADE ANDALUSITE IS IMPORTANT in making high-quality refractories. This material is used in critical areas of the iron and steel industries, including blast furnace stoves. Andalusite imparts superior "creep resistance" compared to bauxite refractories. This means the finished product does not expand and contract as much as bauxite refractories. This results in a longer product life and lower blast furnace operating costs. There is no technically equivalent substitute available in the United States at a cost comparable to andalusite. Bauxite refractories can be used as a substitute but only with inferior results, a reduced product life, and higher operating costs.

The United States imported an average 5,250 tons of andalusite per year for 1984-85 from South Africa. This amount was 100 percent of US imports and consumption, making the United States totally dependent upon South Africa for this material.

France is the only other significant producer of andalusite; however, French andalusite is a different particle size and is technically equivalent to a medium-grade bauxite refractory. US industry has informed the Bureau of Mines that they believe they would lose market share of blast furnace steel production to foreign competition if they were denied supplies of South African andalusite, since they would not be able to offer a technically equivalent, cost-competitive product compared to foreign-made steels. There is no alternate source of material which is technically equivalent and available at a comparable price to substitute for South African andalusite in the United States.

Antimony

Antimony is used as a metal in storage batteries, power transmission equipment, type metal, solder, and ammunition. It is alloyed with other metals to enhance hardness, corrosion resistance and fatigue strength. Its principal non-metallic application (as antimony oxide) is in flame retardants, but antimony compounds are also used as stabilizers in plastics and pigments in paints. Other materials can be substituted for antimony but generally entail cost disadvantages and lead to inconveniences such as the necessity to alter manufacturing processes.

Although US mines produce a small amount of antimony, the United States relied on imports for an average of 59 percent of apparent consumption in 1983-85. China is the largest source of antimony metal imports (48 percent) and Bolivia is the largest supplier of ores and concentrates (45 percent). South Africa supplies only 9 percent of ores and concentrates. South Africa, however, is the main source of antimony oxide imports, accounting for 36 percent of the US total. The value of average US imports of antimony oxide (10,768 short tons, antimony content worth $19.2 million) exceeds that of all other forms of antimony combined (5,686 st worth $9.2 million). Thus, the true extent of US dependence on South Africa for antimony imports is probably greater than the 14 percent of consumption indicated in the statistics because of the relative amount of oxides imported from South Africa.

Alternative suppliers of antimony include the People's Republic of China, the USSR, Bolivia and Mexico. Mexico, however, is not a large supplier. Bolivia has the potential to supply much of US import needs, but its production is mostly from medium and small mines and in recent years has been declining due to internal problems.

Chrysotile Asbestos

Spinning-grade, low-iron chrysotile asbestos has characteristics that make it uniquely attractive to certain defense applications. Its very high strength-to-weight and strength-under-high-temperature qualities are applicable to rocket and missile construction; its electrical and thermal conductivity qualities are applicable to submarine construction. There are no known substitutes that can meet the strength, chemical inertness, and durability characteristics of this grade of asbestos for these applications.

At present, Zimbabwe is the only source of this grade asbestos in the world. It is possible that Canada could resume mining this grade at some time in the future, and it is also possible that South Africa could produce this grade, but the latter cannot be documented. Zimbabwe is heavily dependent on South Africa for the export of this asbestos. Only a very small portion is shipped via Mozambique. The asbestos is produced by a UK subsidiary, but there is no documentation on whether or not South African parastatals are involved in the export or marketing of the product. Because the majority of Zimbabwe asbestos is exported via the South African ports and transportation system, the involvement of South African parastatals in the marketing or export of the product must be considered a present or future possibility. This commodity is therefore placed on the exceptions list. Supporting statistics are not available for this grade of asbestos.

Chromium

Chromium is a metal used in widely diversified products such as stainless, tool, and alloy steels, heat and corrosion-resistant materials, special purpose alloys, alloy cast iron, pigments, metal-plating, leather tanning, chemicals, and refractories for metallurgical furnaces. It is essential to stainless steel and superalloy production, and therefore is vitally important in the defense, aerospace, chemical, power-generation, and transportation industries.

The US produces no chromite ore, and relied on imports for 84 percent of chromium consumption during 1983-85. The remaining 16 percent was supplied through stainless steel scrap recovery and recycling. South Africa, with 34 percent of world chromium ore production in 1985, supplied an average of 73 percent of US imports of chromium and 60 percent of chromium contained in ferrochromium and chromium metal. The United States is also indirectly dependent on South Africa for even more of its chromium, as the second largest supplier of US imports, Zimbabwe, must ship most of its product out through South Africa using South Africa's transportation and port facilities.

The principal alternative supplier is the USSR, which in 1985 accounted for 29.6 percent of total world chromite production and about 15.0 percent gross weight of ferrochromium production. Surplus chromite capacity of the market economy countries in 1984 (outside of South Africa and Zimbabwe) was equivalent to only 43 percent of South Africa's production in that year.

Cobalt

Cobalt is considered critical because of its uses (1) as a constituent in heat- and corrosion-resistant superalloys used in jet engines, (2) in magnetic alloys, (3) as a desulfurizing catalyst in refining crude oil, and (4) as a binder in tungsten carbide for metal cutting and forming, wear resistant applications, and in mining tools. Various materials may be used as substitutes for cobalt in some uses but generally only with a loss of effectiveness. Nickel may be substituted for cobalt without loss of effectivenness in superalloys. No satisfactory substitutes have been found for cobalt as a binder in carbides.

The United States depends on imports to supply usually about 95 percent of its total cobalt consumption, recycled material supplies the remainder. Zaire and Zambia together supplied 58 percent of those imports in 1985. Under current circumstances nearly all cobalt produced in and shipped from Zaire and Zambia must pass through South Africa by truck or rail to the ports of Port Elizabeth or East London. Thus the United States is indirectly dependent on South Africa, by virtue of its transportation system, for supplies of cobalt from southern Africa. Although very little South African cobalt is produced, the involvement of South African parastatals in the marketing or export of the Zairian and Zambian cobalt must be considered a possibility.

Possible alternative suppliers of refined cobalt include Canada, Norway, Japan, and Finland. However, refinery capacity in market economy countries in 1985 (other than Zaire and Zambia) was 36 percent of that of Zaire and Zambia. It was also by-product in nature, as would be other potential sources of new capacity.

Industrial Diamonds (Natural)

The principal uses of industrial diamond stones are in drilling bits and reaming shells, single- or multiple-point diamond tools, diamond saws, diamond wheels, abrasives and diamond wire-drawing dies. Miscellaneous uses include: engraving points, glass cutters, bearings, surgical instruments and special tools. Although non-diamond substitute materials are available for many applications, they are generally not as efficient, cost-effective or as adaptable to the profusion of diamond end uses; and they are less suitable for use in highly automated operations that must perform repetitive functions over long periods within close dimensional tolerances. No efficient substitutes are available for core drilling in hard-rock formations or in concrete.

The United States is totally dependent upon foreign sources for natural industrial diamonds and imports approximately 35 percent of total world production. South Africa supplied approximately 42 percent of US imports from 1983-85.

South Africa is the world's fifth largest producer of industrial diamonds and it is estimated it will supply 12 percent of world demand in 1986. Other important producers and their estimated 1986 share of world production are: Zaire (33 percent), Botswana (12 percent), Australia (27 percent), and the USSR (12 percent). The five largest producers control 95 percent of the market. While South Africa is the fifth largest producer of diamonds, its influence on the market is greater since a private South African company, De Beers' Central Selling Organization (CSO), markets 80-85 percent of all diamond production. Because of the CSO, it is difficult to determine the country of origin of diamonds.

Manganese

The importance of manganese stems from the functions it performs as a desulfurizing, deoxidizing, and alloying element and from its chemical properties. Manganese is essential to the production of virtually all steel, and it is important in the production of cast iron. Manganese increases the strength, toughness, hardness and hardenability of steel, and inhibits formation of embrittling grain boundary carbides. Manganesc is used by the steel industry primarily in the form of ferromanganese or silicomanganese, but some metal is also used. The common dry cell battery uses manganese dioxide, as do glassmaking and the chemical industry. Manganese metal is used as an addition to steel and other alloys in order to introduce manganese without adding impurities. It is used principally in stainless, carbon and alloy steels. Manganese is also added to aluminum in making can stock to impart strength and workability. There is no satisfactory substitute for manganese in its major applications.

The United States is 100 percent dependent on imports for manganese feed material. South Africa supplies about 32 percent of total manganese content in all forms. South Africa supplies 10 percent of US consumption of metallurgical grade ore, which is met entirely by imports. The United States receives 98 percent of metal imports from South Africa which average 38 percent of annual US consumption. South Africa also supplies some 36 percent of manganese ferroalloys (ferromanganese and silicomanganese) which is about 25 percent of total US consumption. Between 80 to 90 percent of manganese ferroalloy demand is supplied by imports.

Alternate sources of manganese ore are available at the current time to replace supplies from South Africa. However, should other countries decide to embargo South African manganese ore, it would create a shortage of ore on a worldwide basis. Alternative supplies of manganese metal would not be available to replace a cutoff of South African metal. Only three countries produce manganese metal, South Africa, Japan and the United States. There is excess manganese metal production capacity in the United States; however, current estimates are that it could replace only 20 percent of South African imports. Japan should not be counted upon to make up a shortfall because it is unclear what excess capacity, if any, is available there. Excess ferroalloy capacity appears to be available in the United States to replace South African ferroalloy imports; however, this capacity has not been in use for several years, and it is difficult to predict how serviceable it is. In any event, US capacity is not price competitive with imports. Any price increase would have a negative impact on the US steel industry, which is already in a poor financial condition. Excess manganese ferroalloy capacity exists outside the United States to replace South African imports. However, there would be a shortage if other countries embargoed South African material. In addition, it is doubtful that a sufficient quantity of manganese feed stock could be made available to ferroalloy producers without ore from South Africa. For the above reasons, all forms of manganese are included in the exemption of the ban on South African parastatal imports.

Platinum-Group Metals

The platinum-group metals have certain extraordinary physical and chemical properties—refractories, chemical inertness, and excellent catalytic activity—which are difficult or impossible to find in other materials. In the United States the largest demand component is for catalysts required by the automobile industry, principally to meet environmental emission standards imposed by US legislation. The chemical industry uses PGM catalysts to manufacture basic industrial chemicals, principally nitric acid. Domestic demand is increasing steadily; and demand in Europe is expected to increase strongly in the future as environmental standards there become stricter. There are substitutes for PGMs in electrical and electronic uses, although those substitutes are considered to be more expensive. Given current technology, there are no known substitutes for PGMs that can approach PGM properties for catalytic uses in the automobile, chemical, and petroleum industries. PGMs are essential and needed by the automotive industry to meet requirements of the US Clean Air Act.

The United States has some limited reserves of PGMs but they are estimated to be equal to less than 3 percent of forecast US cumulative demand between 1983-2000. Nearly all of the world's supplies of PGMs are located in three countries: The Republic of South Africa, the USSR, and Canada. Of these three, only South Africa can produce PGMs as a primary product. Both the USSR and Canada can only produce it as a by-product. South Africa currently produces slightly less than half of total world PGMs and sells nearly all of it to market economy countries

(MEC). The USSR also produces slightly less than half of total world PGMs but sells only about 42 percent of its production to MECs. A ban on South African PGMs would increase US dependence on the USSR assuming that the USSR could give up some of the PGM it now consumes.

About 7 percent of US consumption is supplied domestically through recycling or reclaiming. More than 90 percent of US demand is supplied by imports. In 1985, South Africa directly supplied 46 percent of those imports. The USSR directly supplied 8 percent, and Canada 7 percent. The rest of US imports came from sources that had to acquire most or all of their material from one of the three primary producers. Alternative sources of supply for PGM besides these three producers do not exist in the short term. In the long term, new mines may open in the coming decades in Canada and Australia, where new exploration projects show some promise.

Rutile and Titanium-Bearing Slag

Rutile is essentially crystalline titanium dioxide ($Ti0_2$), and it is the preferred titaniferous raw material for the manufacture of titanium metal because of its high grade and cost effectiveness. It is also the preferable input material for chloride-process $Ti0_2$ pigment manufacturing, which is the desired pigment process because of environmental (the chlorine can be recycled) and cost considerations. Rutile substitutes, including Richards Bay slag (85 percent $Ti0_2$) from South Africa and synthetic rutile from other sources, are made from ilmenite, an iron-titanium oxide ($FeTi0_2$), which is much more abundant than rutile. About one-half of the Richards Bay slag is produced in a suitable size range to be used as a substitute for natural rutile. The other half of Richards Bay slag can substitute for ilmenite in the production of sulfate-process pigment.

In 1985, about one-third of US total consumption $Ti0_2$ in concentrates came from rutile and rutile substitutes; the other two-thirds was from ilmenite and from titanium slag that was not usable as a rutile substitute. Of US consumption of rutile and rutile substitutes in 1985, 84 percent was used for manufacture of $Ti0_2$ pigments, and 16 percent for titanium metal. US titanium metal production is derived entirely from rutile or rutile substitutes such as Richards Bay slag. The largest US market (48 percent) for TiO_2 pigment is for use in surface coatings. Because of its high refractive index, it imparts whiteness, opacity, and brightness. About 22 percent of TiO_2 pigment was consumed as paper coatings to improve opacity, brightness, and printability. This pigment is also used in plastics and many miscellaneous applications. Titanium metal is used in aerospace applications in engines and airframes because of its high strength-to-weight ratio and resistance to heat and corrosion. There is no satisfactory substitute for titanium in aerospace applications. Also, at present there is no cost-effective substitute for TiO_2 in pigments because there is no other material which is available in sufficient quantities that has properties similar to TiO_2 (high refractive index, dispersion qualities, etc.).

The United States imported 123,000 short tons of TiO_2 in rutile and titanium-bearing slag from South Africa in 1985. This was 18 percent of total imports of titanium concentrates and 13 percent of US consumption. However, the United States was even more dependent upon South Africa for its supply of rutile and rutile substitutes, including the half of imported Richards Bay slag that is a suitable size to be a rutile substitute. In 1985 the United States imported 82,000 tons of TiO_2 in rutile and rutile substitutes (41,000 tons rutile; 41,000 tons Richards Bay slag) from South Africa. This was 39 percent of total imports and 26 percent of total US consumption of these materials. In considering the total impact of imports from South Africa of titanium raw materials on the US economy, all of the Richards Bay slag is important because of its high titanium content.

Australia and Sierra Leone are the major alternate sources of rutile. However, additional quantities to make up a loss of South African supplies are not likely to be available from these countries because of the record-high world demand for titanium concentrates for the production of titanium dioxide pigments. Sufficient supplies of titanium would be available for metal production, since it accounts for only five percent of total consumption. Loss of South Africa as a source of titanium concentrates would create severe shortages in the American pigment industry, which has been operating at close to full capacity for the past three years to satisfy growing demand. For aerospace applications, there is no satisfactory substitute. Also, there is no cost-effective substitute for TiO_2 pigment.

Vanadium

Vanadium is used mainly as an alloying agent in steel and cast iron production. It imparts high temperature abrasion resistance to die steels used for high speed tools and turbine rotors. To structural steels it adds increased toughness, ductility and strength, and is therefore important to the automobile industry and the construction of bridges, offshore drilling platforms, oil pipelines, etc. About 15 percent of the vanadium used in the US goes into titanium alloys for use in aircraft and related industries. Vanadium alloys are also used as catalysts in the chemical industry, largely in the production of sulfuric acid.

Various metals such as columbium, manganese, titanium, chromium, molybdenum and tungsten can substitute for vanadium in steel applications, but all have some technical and/or economic drawbacks. The PGMs can substitute for vanadium compounds in some catalytic applications, but they are more costly and more susceptible to poisoning in sulfuric acid production. There is no acceptable substitute for vanadium in titanium alloys.

The United States produced 54 percent of its domestic vanadium needs in 1984, primarily as a by-product or co-product of uranium. Forty-six percent of US consumption was imported, and about 40 percent of US imports came from South Africa. From a solely strategic point of view, the US could probably supply all of its own vanadium needs. However, to make US production viable, significant price rises would be necessary. In fact, based on world prices and US production costs in recent years, the Secretary of Energy made a formal finding in September 1985, that the US uranium/vanadium mining and milling industries were already not viable. In 1986 100 percent of US imports for consumption of vanadium-bearing slags and chemicals were from South Africa.

The OECD countries are highly dependent on imports of vanadium pentoxide and vanadium-bearing slags from South Africa. South Africa was the world's second largest producer (after the USSR) of vanadium ores in 1983 and accounted for about 60 percent of the vanadium mined by the market economy countries. Between 1980 and 1983, 44 percent of US vanadium imports came directly from South Africa. Additional material was processed in Western Europe and then imported as ferrovanadium or pentoxide. South Africa indirectly accounts for 100 percent of US ferrovanadium imports for consumption. Japan, the Federal Republic of Germany and France are more heavily dependent on South African imports than the United States. Any extended cutback in supplies from South Africa would seriously upset the balance between supply and demand in the market economy countries. Sizable, near-term deficits could only be offset by material from China or the USSR. The permanent closure of Finland's two vanadium mines in 1985 has made the United States even more dependent on South African vanadium.

Source: Department of State, *CERTIFICATION*, 7 January 1987,
(Signed John C. Whitehead)

Appendix C
Mobilization Plans

Note. This excerpt of a letter from William H. Taft, IV, Deputy Secretary of Defense, 17 March 1987, to the Honorable Jim Wright, Speaker of the House of Representatives, lists Reagan administration policy on Emergency Mobilization Preparedness and Department of Defense mobilization plans in response to administration policy.

THE REAGAN ADMINISTRATION'S NATIONAL SECURITY strategy placed emphasis on the twin objectives of (1) restoring the nuclear balance and (2) being able to fight a global conventional war of prolonged duration.

NSDD 47: Emergency Mobilization Preparedness

In response to the need to place greater reliance on war potential (mobilization power), the President issued NSDD-47. This Presidential directive entitled "Emergency Mobilization Preparedness" states

> It is the policy of the United States to have an emergency mobilization preparedness capability that will ensure that government at all levels, in partnership with the private sector and the American people, can respond decisively and effectively to any major national emergency with defense of the United States as the first priority.

As regards military mobilization, NSDD-47 specified that the United States mobilization plans and programs will be designed to increase our capabilities to

—expand the size of the force from partial through full to total mobilization;

—deploy forces to theaters of operations, and sustain them in protracted conflict; and

—provide military assistance to civil authority, consistent with national defense priorities and applicable legal guidelines.

Sec. 3206. Report on War Emergency Situations and Mobilization Requirements

This section requires the Secretary of Defense to submit to Congress a report describing war emergency situations that would necessitate *total mobilization* of the economy of the United States for a conventional global war. In response to the specific questions posed by the Congress the following estimates are provided:

1. *The Length and Intensity of the Assumed Emergency*

a. Any war involving United States and Soviet forces in combat carries with it the potential of becoming a world war and, therefore, will necessitate total mobilization (i.e., expansion of the current force).

b. Such a war might begin as a localized conventional conflict.

c. Under these conditions, the United States would attempt to keep the war contained to the theater of origin, win it, or otherwise end it on terms acceptable to the US. In any case, our long term objective is to be able to sustain conventional combat for as long as the enemy can sustain this mode of warfare.

d. We would also attempt to protect our vital interests at the lowest level of violence that is consistent with this goal.

e. Thus, the length and intensity of such a war cannot be predetermined. However, once begun, the mobilization process will continue at whatever level of effort is necessary to assure a rapid and successful outcome.

2. *The Military Force Structure to be Mobilized.*

a. The United States mobilization system/process is designed to reinforce deterrence in a crisis and/or sustain combat if deterrence should fail. As regards the military force structure to be mobilized the following points are relevant:

(1) In comparison to pre-World War II force levels, we have in being a relatively large, well trained and equipped, force—about five million in all, including active duty Service members,

civilian employees, and Reserve forces. These land, sea, and air forces are forward deployed at 335 overseas installations in 21 countries and 19 installations in six United States territories. In contrast, the mobilized strength of the armed forces in 1940 was about one-half million, and only 75,000 troops were forward deployed overseas in defense of United States territories.

(2) Therefore, we have developed mobilization plans and programs to bring the peacetime force and war reserve stocks to wartime authorized levels (partial to full mobilization) and to expand the force as necessary. In this regard, the Secretary of Defense said that we and our allies must be prepared to respond to warning indicators that are highly ambiguous, decided upon quickly, sustained—if necessary, for a prolonged period—until the ambiguity is resolved, and repeated every time the warning indicators demand it. A policy that provides for such responses, as a routine procedure, can help to avert crises and strengthen deterrence. By contrast, being prepared to respond only to warning that is unambiguous means being prepared for the kind of warning we are least likely to get.

b. In response to this policy we have:

(1) Developed a critical items list of munitions and equipment which the Commanders-in-Chief (CINCs) of the various theaters have determined to be "mission-decisive." The Services and the Joint Staff, in turn, have conducted production base analyses to determine industrial capacity to surge-produce these critical items and have identified steps (e.g., the purchase of long lead time components) which can be taken in peacetime to increase the output of these critical items in *response to warning*.

(2) The Services have determined the requirements to bring the current force to wartime levels of readiness (i.e., C3) and sustainability (partial and full mobilization requirements).

(3) The JCS have estimated the capability of the industrial base to produce the planning force in three years under full industrial mobilization. Certain long lead items (e.g., ships) cannot be built in three years.

c. The estimated cost to build and maintain the planning force under the three-year total mobilization scenario is approximately $1 trillion above current five-year defense plan levels.

3. *Department of Defense Needs for the National Defense Stockpile.*

Global wars, such as those experienced in World War I and II, would undoubtedly be conducted differently in the future due to revised strategies, tactics, force levels, technologies, weapon systems, and the evolutionary changes in the United States industrial base. Therefore, DoD concerns regarding our industrial surge and mobilization capabilities go well beyond the stockpiling of critical mineral ores. Above all else, we need increased production surge capability, especially during the early stages of a national emergency. To this end, we look to industrial prepardness programs for increasing industrial capacity and responsiveness to overcome the shortages and bottlenecks that are expected. Therefore, as proposed by the President, national stockpile planning for strategic and critical materials must be updated and synchronized with military strategy, technological changes, and industrial modernization to meet DoD needs.

Appendix D
National Defense Stockpile Policy: 1985 White House Release

Background

THE PRESIDENT HAS DECIDED TO PROPOSE A MODERNIZATION of the National Defense Stockpile of strategic materials. This proposal comes after 2 years of interagency study and thousands of hours of review at the staff and policy levels at twelve different agencies. The Administration intends to consult and work with the Congress on this important national security program before the new stockpile goals are transmitted.

The National Defense Stockpile is a reserve of nonfuel materials that the United States would require in a conflict, but that might not be available in sufficient quantities from domestic or reliable foreign sources. The previous Administration in 1979 calculated the United States' stockpile needs to be $16.3 billion for 62 materials using May 1985 prices. Toward this goal, the stockpile contains $6.6 billion in materials. The USG possesses an additional $3.5 billion of materials that are surplus to our requirements under the 1979 goals. Thus, unmet materials needs are $9.7 billion under the 1979 goals.

The President's April 5, 1982, "National Materials and Minerals Program Plan and Report to Congress" announced "a major interdepartmental effort to improve the Nation's preparedness for national mobilization." Part of the review was to address the potential national security impacts of shortages of strategic and critical materials. The review covered the 42 most significant materials in the stockpile. The remaining materials will be reviewed at a later date.

The key elements of the Nation's stockpile policy are as follows:

—The National Defense Stockpile will be sufficient to meet the military, industrial and essential civilian needs for a 3-year conventional global military conflict, as mandated by Congress in 1979.

—The conflict scenario used is to be consistent with the scenarios developed by DOD.

—The stockpile should reflect detailed analyses regarding the conflict period: essential civilian, industrial and defense mobilization requirements, foreign trade patterns, shipping losses, petroleum availability, and foreign and domestic demand and production levels for the materials in question.

Policy Decisions

On the basis of the new stockpile study of materials requirements and supplies during a protracted military conflict, the President has decided that the stockpile for the 42 materials studied will now contain $6.7 billion in materials and include two tiers. Goals of $.7 billion (Tier I) are proposed for materials that would be required during a protracted military conflict that would not be available in sufficient quantities from domestic or reliable foreign sources. The stockpile also will contain a Supplemental Reserve of strategic and critical materials currently valued at $6 billion (Tier II). The Supplemental Reserve will contain materials that the USG already possesses. This reserve will offer additional assurance against materials shortages during a period of military conflict. Both Tiers of stockpile provide over one year's peacetime levels of imports for such materials as chromium, manganese, cobalt and tantalum. These new stockpile goals will eliminate the $9.7 billion unmet goal.

The new stockpile will result in surplus materials of $3.2 billion, as opposed to the $3.5 billion surplus calculated by the previous Administration. The mix of materials considered to be surplus, however, is different.

The President has decided to sell a portion ($2.5 billion out of $3.2 billion) of the surplus materials stocks in a manner—over the next five years—that minimizes market impacts. An interagency group will evaluate ways to ensure that stockpile sales do not cause undue market disruption.

Receipts from the sales program will go to fill unmet materials goals under the 1984 study, including any goals that may result from analyses of the twenty materials yet to be studied, including any new, high-technology materials; the remainder will go to reduce the deficit. The stockpile goals planning assumptions also will be used for other appropriate mobilization preparedness areas.

Study Process

The 1984 stockpile study completed by the Administration included a review of the analysis, methods and assumptions used by the previous Administration in the 1979 study. This review concluded that a number of basic errors and unrealistic assumptions were used in the 1979 study. The present study relied on more realistic assumptions regarding oil availability, essential civilian requirements and domestic materials production.

The new stockpile, unlike the one proposed in 1979, does not reflect the stockpiling of materials to ensure non-essential consumer production in a protracted military conflict. The stockpile does reflect essential civilian goods production with per capita consumption at more than twice the WWII level.

In the 1984 study, substantial improvements were made in analytic methods for estimating material requirements and available supply. These changes, the correction of errors and the use of more plausible assumptions, are the primary reasons for the revised goals. The 1984 study was started in 1983 and relied on actual data up to and including 1982 for all phases of the analysis. In all areas, the latest, best available data was used. By contrast, the previous 1979 stockpile goals relied on 1967 data in many cases.

Notes

Chapter 1: Mineral Dependency: The Vulnerability Issue

1. Stanley H. Dempsey, "Environmental Health and Safety Regulations," *Mining Congress Journal*, January 1980, p. 4.

2. US, Department of Interior, *Mineral Commodity Summaries 1987* (Washington, DC: Government Printing Office, 1987), p. 1 facing.

3. L. Harold Bullis and James E. Mielke, *Strategic and Critical Materials* (Boulder, Colorado: Westview Press, 1985), p. 27.

4. Ibid., p. 54.

5. US, Bureau of Mines, *Minerals Yearbook*, Vol. 3, 1984, p. 850.

6. Ibid., p. 845.

7. "Science Behind the News," *Discover*, September 1986, p. 14.

8. *Mineral Commodity Summaries, 1987*, pp. 35, 99.

9. *Strategic and Critical Materials*, p. 54.

10. Ibid., p. 50.

11. US, Office of Technology Assessment, *Strategic Materials: Technologies to Reduce US Imports Vulnerability*, Washington, DC, May 1985, pp. 165-66.

12. *Strategic and Critical Materials*, p. 49.

13. D. E. Meadows et al., *The Limits to Growth*, 2d ed. (New York: University Books, 1974).

14. *Strategic and Critical Materials*, pp. 101-2.

15. Ibid., pp. 106-9.

16. US, Library of Congress, Congressional Research Service, *A Congressional Handbook on US Materials Import Dependency/Vulnerability, A Report to the House Sub-committee on*

Economic Stabilization of the Committee on Banking, Finance, and Urban Affairs, September 1981, pp. 37-38.

17. Parastatals are private companies in South Africa that are subject to some degree of control by the central government. In the case of most minerals, the South African Chamber of Mines is involved.

18. Certification, John C. Whitehead, Department of State, 7 January 1987.

19. Robert England, "The Critical Four," *Washington Times*, 30 September 1985, p. 9C.

20. The platinum group comprises six metals: platinum, palladium, rhodium, iridium, osmium, and ruthenium.

21. See "To Thwart Pretoria, Mugabe Seeks to Revive Africa's Rails," *Washington Times*, 10 April 1987, p. 9A; "Angola Rebels in Offer on Rail Link," *New York Times*, 27 March 1987, p. A3; "US Pledges Aid to South Africa's Neighbors," *New York Times*, 6 February 1987, p. A5.

22. "Front-line Rail Monte Debated," *Washington Post*, 1 January 1987, p. A23.

23. "US Africa Aid Plan Appears to Offer Less Than Anticipated," *Washington Post*, 5 February 1987, p. A12.

24. The amounts of these metals actually contained in the engine are considerably less due to machining and forming losses.

25. *Strategic Materials*, pp. 281-82.

26. Federal Emergency Management Agency, *Stockpile Report to the Congress*, FEMA 36, October 1986, p. 10 and p. 54.

Chapter 2: A Key Distinction: Dependency Versus Vulnerability

1. David Owen, *Metal Bulletin*, 25 October 1985.

2. Steve Salerno, *American Legion Magazine*, April 1986, p. 43.

3. US, Congress, House, Committee on Science and Technology, *Emerging Issues in Science and Technology: A Compilation of Reports on CRS Workshops*, 96th Cong., 2d sess., December 1980, p. 12.

4. See, for example, Emery N. Castle and Kent A. Price, *US Interests and Global Natural Resources*, Resources for the Future, Inc., Washington, DC, pp. 58-64.

5. US, Office of Technology Assessment, *Strategic Materials: Technologies to Reduce US Import Vulnerability*, OTA-ITE-248, May 1985, p. 93.

6. *Report of the National Commission on Supplies and Shortages, Government and the Nation's Resources*, December 1976, pp. 30-31.

7. *US Interests and Global Natural Resources*, p. 58.

8. *Strategic Materials*, p. 93.

9. "All That Glitters is Not US Gold," *Washington Post*, 28 April 1987.

10. *Strategic Materials*, p. 222.

11. Ibid., p. 223.

12. Ibid., p. 239.

13. US, Office of Technology Assessment, *Effects of the 1978 Katanganese Rebellion on the World Cobalt Market, Final Report to the Office of Technology Assessment*, December 1982, pp. 1-6.

14. The "buy-to-fly" ratio is the ratio of input material to the amount contained in the final fabricated part.

15. *Strategic Materials: Technologies to Reduce US Import Vulnerability*, pp. 227-38.

16. US, Bureau of Mines, *Mineral Facts and Problems*, 1985 ed.,Bureau of Mines Bulletin 675, p. 495.

17. Ibid., 496.

18. L. Harold Bullis and James E. Mielke, *Strategic and Critical Materials* (Boulder, Colorado: Westview Press, 1985), p. 193.

19. Ibid., p. 197.

20. *Strategic Materials*, p. 266.

21. National Materials Advisory Board, National Research Council, *Basic and Strategic Metals Industries: Threats and Opportunities*, NMAB-425 (Washington, DC: National Academy Press, 1985), p. 109.

22. Ibid., p. 121.

23. Ibid., pp. 118-19.

24. Letter from Dwight D. Eisenhower to Senator Clifford P. Case, 24 September 1963.

25. US, Congress, House, *Hearing on H.R. 3743 Before the Seapower and Strategic and Critical Materials Subcommittee of the House Committee on Armed Services*, Simon D. Strauss, H.A.S.C. No. 99-29, 27 February 1986, p. 31.

26. Timothy D. Gill, *Industrial Preparedness* (Washington, DC: US Government Printing Office, 1984), p. 3.

27. Ibid., p. 6.

28. Ibid., p. 8.

29. US, Library of Congress, Congressional Research Service, *The Reagan Administration Proposes Dramatic Changes to National Defense Stockpile Goals*, Alfred R. Greenwood, 86-578 ENR, Washington, DC, 8 February 1986, p. 33.

30. Simon Strauss, p. 31.

31. Timothy D. Gill, p. 12.

32. Ibid., p. 16.

33. Ibid., p. 17.

34. Robert T. Jordan, "Most of It From Overseas," *Sea Power*, December 1985, p. 51.

35. Alfred R. Greenwood, p. 35.

36. Robert T. Jordan, "Most of It From Overseas," p. 51.

37. Alexander B. Trowbridge, *Hearing on H.R. 3743*, p. 31.

38. Ibid., p. 48.

39. Robert T. Jordan, "Most of It From Overseas," p. 51.

40. US, Congress, House, *Testimony by Rear Admiral Robert J. Hanks, USN (Ret.), to Subcommittee on Seapower and Strategic and Critical Materials*, 18 March 1987, p. 9.

41. Timothy D. Gill, pp. 31-32.

42. Alfred R. Greenwood, p. 4.

43. From a speech given before the Joint Industry-Government Telecommunications Industry Mobilization Group by John D. Morgan, Chief Staff Officer, US Bureau of Mines, 16 December 1986.

44. US, Department of Defense, *The Effects of a Loss of Domestic Ferroalloy Capacity*, Myron G. Myers, Donna J. S. Peterson, and Robert L. Arnberg, Logistics Management Institute,

Bethesda, MD, June 1986, (Department of Defense Contract MDA 903-85-C-0139).

45. US, Bureau of Mines, *South Africa and Critical Materials*, Open File Report 76-86, July 1986, p. 50.

46. Ibid., p. 55.

47. *OTA Report*, p. 163.

48. Ibid., p. 169.

49. This assessment is calculated on the basis of chromium contained both in ores and ferroalloys where appropriate.

50. *The Effects of a Loss of Domestic Ferroalloy Capacity*, pp. 3-16.

51. US, Bureau of Mines, *Mineral Commodity Summaries 1987*, 1987, p. 34.

52. *Platinum, 1986 Interim Review*, Johnson Matthey, London, England, November 1986, p. 9.

53. *OTA Report*, pp. 239 and 252.

54. The rare earths, also called lanthanides, are a group of 15 chemically similar elements with atomic members 57 through 71. The major producers are the United States, Australia, China, and India.

55. *South Africa and Critical Materials*, p. 50.

Chapter 3: Strategic Minerals Policy and Public Good

1. The author is indebted to Joe Goldberg, who is on the staff of the National Defense University, for articulating the concept of the public good. This section of the text makes heavy use of both his ideas and his verbiage.

2. US, General Accounting Office, *National Defense Stockpile: National Security Council Study Inadequate to Set Stockpile Goals*, GAO/NSIAD-87-146 (Washington, DC: Government Printing Office, May 1987).

3. For a discussion of the history of stockpiling, see L. Harold Bullis and James L. Mielke, *Strategic and Critical Materials* (Boulder, Colorado: Westview Press, 1955), pp. 208-24.

4. US, Department of Defense, *Strategic and Critical Materials Report to the Congress* (Washington, DC: US Government Printing Office, 1988), p. 4.

5. Office of the Press Secretary, The White House, 8 July 1985.

6. US Congress, House, Congressional Research Service, *A Congressional Handbook on US Materials Import Dependency/Vulnerability, Report to the Subcommittee on Economic Stabilization of the Committee on Banking, Finance, and Urban Affairs* (Washington, DC: Government Printing Office, September 1981), pp. 331-32.

7. Ibid., p. 342.

8. Ibid., p. 343.

9. John D. Morgan, Chief Staff Officer, US Bureau of Mines paper presented for 5th Annual Mobilization Conference, Industrial College of the Armed Forces, Fort McNair, Washington, DC, 22-23 May 1986.

10. US, Bureau of Mines, *Mineral Commodity Summaries, 1986*, p. 3; *Mineral Commodity Summaries, 1989*, pp. 4-5.

11. *Mineral Commodity Summaries,1986*, p. 1 (facing).

12. US, Office of Technology Assessment, *Management of Fuel and Non-fuel Minerals in Federal Land, Current Status and Issues*, (Washington, DC: Government Printing Office, 1979), p. 217.

13. *National Strategic Materials and Minerals Program Advisory Committee Assessment*, May 1986, p. 5.

14. US, Office of Technology Assessment, *Strategic Materials: Technologies to Reduce US Import Vulnerability* (Washington, DC: Government Printing Office, May 1985), pp. 201-4.

15. Ibid., pp. 150, 167, and 185.

16. Bullis and Mielke, *Strategic and Critical Materials*, p. 209.

17. US, General Accounting Office, *National Defense Stockpile: Adequacy of National Security Council Study for Setting Stockpile Goals*, GAO/NSIAD-86-177BR (Washington, DC: Government Printing Office, August 1986).

18. See *Hearings before the Subcommittee on Seapower and Strategic and Critical Materials*, 18 March 1987.

19. Hans H. Landsberg, "Stockpiles Are Inadequate," *Wall Street Journal*, 14 August 1986.

20. "Resource Wars: The Myth of American Mineral Vulnerability," *Defense Monitor*, Vol. 14, No. 9, 1985, p. 2.

21. Industrial Mobilization for Defense, Agreement between the United States of America and Canada, 26 October 1950.

22. US, Bureau of Mines, *Mineral Commodities Summaries, 1989* (Washington, DC: Government Printing Office, 1989), p. 7; Department of Defense, *The Effects of a Loss of Domestic Ferroalloy Capacity*, Logistics Management Institute, Bethesda, Maryland, June 1986 (Department of Defense Contract MDA 903-85-C-0139).

23. The ensuing discussion is from Bullis and Mielke, *Strategic and Critical Materials*, pp. 95-99.

24. US, General Accounting Office, *Federal Encouragement of Mining Investment in Developing Countries for Strategic and Critical Minerals Has Been Only Marginally Effective*, Washington, DC, 1982.

25. US, Office of Technology Assessment, *Strategic Materials: Technologies to Reduce US Import Vulnerability*, p. 322.

Chapter 5: In the Final Analysis

1. Hardy L. Merrit and Luther F. Carter, eds., *Mobilization and the National Defense* (Washington, DC: Government Printing Office, 1985), p. 22.

The Author

Kenneth Allen Kessel wrote this book while he was a Senior Fellow at the National Defense University. A career intelligence officer since 1969 in Washington, DC, Mr. Kessel specializes in energy and mineral resource economics. He earned his bachelor of science degree at the Illinois Institute of Technology and master of arts at American University. Mr. Kessel resides in Reston, Virginia.

88697210R00093

Made in the USA
San Bernardino, CA
16 September 2018